The Science Behind Common Objects: Grasping the Physics in Our Daily Lives

Shah Rukh

Published by Shah Rukh, 2024.

THE SCIENCE BEHIND COMMON OBJECTS: GRASPING THE PHYSICS IN OUR DAILY LIVES

First edition. May 24, 2024.

Written by Shah Rukh.

Table of Contents

Prologue

In the midst of our busy and routine-filled lives, there lies a concealed realm of scientific knowledge waiting to be unveiled. This realm is the world of physics, where the ordinary items that fill our homes, workplaces, and communities possess a secret language of forces, energies, and principles that govern their behavior and purpose.

Welcome to "The Science Behind Common Objects: Grasping the Physics in Our Daily Lives." Within the pages of this book, we embark on a journey of exploration, delving into the captivating physics that underlies the objects we encounter on a daily basis – from the simple spoon to the sophisticated smartphone, from the familiar light bulb to the ubiquitous door hinge.

As we delve into the scientific explanations behind these commonplace objects, we will uncover the intricate mechanisms, phenomena, and principles that shape their design, functionality, and usefulness. We will unravel the enigmas of motion, energy, and forces that influence their behavior, and in doing so, we will develop a newfound admiration for the intricate and intricate nature of the physical world that envelops us.

However, this book is more than just a scientific expedition; it is a celebration of curiosity, awe, and discovery. It extends an invitation to observe the world around us with fresh perspectives, to marvel at the brilliance of human ingenuity, and to embrace the interconnectedness of science and our everyday lives.

So, let us embark on this journey of exploration and enlightenment together, as we uncover the concealed physics behind the objects that enhance and define our daily experiences. Together, let us unravel the mysteries of the universe, one common object at a time.

Chapter 1: Introduction: The Physics Around Us

Physics is an integral part of our everyday lives, often in ways we don't even notice. From the moment we wake up to the time we go to bed, our daily activities are governed by the principles of physics. Understanding these principles can not only give us a greater appreciation for the world around us but also help us make informed decisions in various aspects of our lives.

Consider the alarm clock that wakes you up in the morning. The clock's ability to measure time accurately is a result of precise physical principles. Many modern alarm clocks use quartz crystals, which vibrate at a consistent frequency when an electric current is applied. This property, known as piezoelectricity, ensures the clock keeps accurate time, down to the second. This concept extends to various timekeeping devices, from wristwatches to global positioning systems (GPS), all of which rely on the consistent, predictable behavior of physical systems.

As you get out of bed, you might walk to the bathroom. Walking itself is a fascinating application of physics. It involves the principles of mechanics, particularly Newton's laws of motion. When you push off the ground with your foot, the ground pushes back with an equal and opposite force, propelling you forward. This is Newton's third law in action. Your muscles and bones work together as a system of levers and pulleys, optimizing the forces to move your body efficiently. The friction between your feet and the floor prevents slipping, allowing you to maintain balance and control your movement.

In the bathroom, you might turn on the light, demonstrating electromagnetism. When you flip the switch, you complete an electrical circuit, allowing current to flow through the wires to the light bulb. In incandescent bulbs, this current heats a tungsten filament until it glows, emitting light. In fluorescent and LED bulbs, different processes convert electrical energy into visible light more efficiently. Fluorescent bulbs use gas and a coating of phosphor inside the tube, while LEDs rely on semiconductor materials that emit light when an electric current passes through them.

Showering involves the physics of fluids and heat transfer. Water flows from the showerhead due to the pressure difference created by the water supply system. The water heater warms the water through conduction, where heat is transferred from the heating element to the water. As the warm water hits your body, it transfers heat to your skin, a process governed by the principles of thermodynamics. The steam rising from the shower illustrates the phase change of water from liquid to gas, an endothermic process that absorbs energy from the surrounding air.

Cooking breakfast involves various physical phenomena. Heating a pan on the stove involves conduction, where heat transfers from the stove burner to the pan. Cooking food in the pan involves convection, where heat circulates through the air or liquid in the pan to evenly cook the food. When you crack an egg into the pan, the proteins in the egg white undergo a process called denaturation, where the proteins unfold and form new bonds, changing from a liquid to a solid state. This is a result of the application of thermal energy breaking and reforming molecular bonds.

Drinking a cup of coffee demonstrates the principles of fluid dynamics and heat transfer. When you pour hot coffee into a cup, the heat from the coffee transfers to the cup and the surrounding air. Over time, the coffee cools, illustrating the second law of thermodynamics, which states that heat energy naturally flows from a hotter object to a cooler one until thermal equilibrium is reached. Stirring the coffee with a spoon introduces fluid dynamics, creating vortices and mixing the liquid, distributing heat and any added substances like sugar or cream more evenly.

Driving to work involves a complex interplay of physics concepts. The car's engine converts chemical energy from fuel into mechanical energy through internal combustion. The transmission system then transfers this energy to the wheels, propelling the car forward. The friction between the tires and the road provides the necessary grip for acceleration, deceleration, and turning. Understanding the principles of aerodynamics helps car manufacturers design vehicles that minimize air resistance, improving fuel efficiency and stability at high speeds. The principles of rotational motion and centripetal force govern the car's ability to navigate curves without skidding off the road.

Even at work, physics is at play. If you use a computer, you're engaging with the principles of electromagnetism and quantum mechanics. The flow of electrons through the computer's circuits and the binary logic that underpins its operations are fundamental concepts in physics. The display screen, whether it's an LCD, LED, or another type, involves the manipulation of light and color to present information. If you print documents, the printer uses principles of electrostatics and mechanics to transfer ink or toner onto paper.

The physics around us is also evident in entertainment and leisure activities. Watching a movie involves understanding the properties of light and sound. Projectors and screens use the principles of optics to display images, while speakers convert electrical signals into sound waves that we can hear. When you play sports, the physics of motion, force, and energy are crucial. Kicking a soccer ball, for example, involves transferring kinetic energy from your foot to the ball, determining its speed and trajectory based on the angle and force of the kick.

Even at rest, physics is present. When you lie down to sleep, your mattress and pillow are designed using principles of material science and mechanics to provide comfort and support. The structural integrity of your home, from the foundation to the roof, is a testament to applied physics, ensuring it can withstand various forces like gravity, wind, and, in some regions, seismic activity.

Understanding the physics around us can deepen our appreciation for everyday objects and activities. It's a reminder that the natural laws governing the universe are constantly at work, shaping our experiences and interactions. From the microcosm of atomic interactions to the macrocosm of celestial mechanics, physics provides a framework for understanding the world in a coherent and predictable way. Embracing this knowledge can inspire curiosity, innovation, and a profound respect for the intricacies of the universe.

Chapter 2: The Science of Sitting: How Chairs Support Us

The science of sitting, especially how chairs support us, involves a fascinating interplay of physics, biology, ergonomics, and materials science. A chair is more than just a piece of furniture; it is a complex structure designed to interact with the human body in ways that promote comfort, stability, and health. Understanding how chairs support us requires a deep dive into the principles that govern their design and function.

When we sit in a chair, our body exerts a force downward due to gravity. This force, our body weight, must be counteracted by an equal and opposite force from the chair to keep us stable and upright. This is a direct application of Newton's third law of motion: for every action, there is an equal and opposite reaction. The chair's structure distributes our body weight across its various components, including the seat, backrest, legs, and armrests. The design and materials used in these components determine how effectively the chair can support us.

The seat of the chair is crucial in distributing the pressure exerted by our body. Ideally, the seat should provide even support across our buttocks and thighs to avoid pressure points that can lead to discomfort or even health issues like pressure sores. To achieve this, the seat is often contoured to match the natural shape of the human body. Ergonomic chairs, in particular, incorporate designs that promote a slight forward tilt of the pelvis, encouraging a natural S-curve in the spine, which helps maintain proper posture and reduces strain on the lower back.

The backrest of a chair plays a significant role in supporting the spine. The human spine has a natural curve, with the lower back (lumbar region) curving inward and the upper back (thoracic region) curving outward. A well-designed chair will have a backrest that provides lumbar support, fitting snugly into the curve of the lower back to maintain this natural posture. This support helps to distribute the weight and reduce the load on the spine and surrounding muscles, preventing slouching and minimizing the risk of back pain.

The height of the chair is another important factor in proper support. The seat height should be adjustable so that the user's feet can rest flat on the floor, with thighs parallel to the ground. This position helps maintain proper circulation in the legs and reduces the risk of developing conditions like deep vein thrombosis (DVT). Additionally, an adjustable seat height allows users of different sizes to use the same chair comfortably, making it a versatile option in environments like offices where multiple people might use the same chair.

Armrests can significantly enhance the support provided by a chair. They offer a place to rest the arms, reducing the load on the shoulders and neck. Properly positioned armrests should allow the user to maintain a relaxed posture, with elbows bent at approximately a right angle and shoulders in a neutral position. Armrests that are too high or too low can cause strain and discomfort, highlighting the importance of adjustability in ergonomic chair design.

The materials used in constructing a chair also influence its ability to support the user. The seat and backrest often incorporate cushioning materials like foam, which provide a balance of comfort and support. Memory foam, for instance, can contour to the shape of the body, distributing weight evenly and reducing pressure points. However, too much cushioning can lead to a lack of support, causing the body to sink too deeply into the chair and resulting in poor posture.

The frame of the chair needs to be strong and stable to safely support the user's weight. Common materials for chair frames include wood, metal, and high-strength plastics. Each material has its advantages:

wood can be aesthetically pleasing and sturdy, metal offers strength and durability, and high-strength plastics can provide flexibility and resilience. The design of the frame must ensure that the chair can withstand regular use without bending, breaking, or losing stability.

One often overlooked aspect of chair design is the importance of mobility and adjustability. Chairs with wheels or casters allow users to move around their workspace easily, reducing the need to stretch or strain to reach different areas. Swivel chairs enable users to turn and shift positions without having to twist their bodies, which can reduce the risk of back and neck strain. Adjustable features, such as seat height, backrest angle, and armrest height, enable users to customize the chair to their specific needs, promoting better posture and comfort.

Beyond the immediate physical support, the design of chairs also considers the long-term health and well-being of users. Prolonged sitting has been linked to various health issues, including obesity, cardiovascular disease, and musculoskeletal disorders. Ergonomic chairs aim to mitigate these risks by promoting movement and encouraging better posture. Features like tilt mechanisms allow users to recline slightly, which can reduce pressure on the spine and encourage blood flow. Some chairs incorporate dynamic sitting principles, where the seat and backrest move in response to the user's movements, promoting active sitting and reducing the static load on the body.

In workplaces and educational settings, the choice of chairs can significantly impact productivity and learning outcomes. Comfortable and supportive chairs can reduce fatigue and discomfort, allowing individuals to focus better and work or study for longer periods. In contrast, poorly designed chairs can lead to discomfort, distraction, and decreased productivity. This understanding has led to the development of specialized seating solutions tailored to specific environments, such as office chairs, classroom chairs, and task chairs.

Chapter 3: The Everyday Spoon: Principles of Leverage

The everyday spoon, a simple yet ubiquitous utensil, is a fascinating example of the principles of leverage in action. Despite its unassuming appearance, the spoon embodies complex physics principles that make it an effective tool for a variety of tasks, from eating soup to measuring ingredients.

Leverage is a mechanical advantage gained by using a lever, a rigid bar that rotates around a fixed point called the fulcrum. The lever amplifies an input force (effort) to provide a greater output force (load) with less effort. This fundamental principle is articulated in the law of the lever, formulated by Archimedes, which states that the ratio of the output force to the input force is equal to the ratio of the lengths of the lever arms on either side of the fulcrum.

There are three classes of levers, each defined by the relative positions of the fulcrum, effort, and load. The everyday spoon can function as an example of all three classes, depending on how it is used.

First-Class Lever:

In a first-class lever, the fulcrum is positioned between the effort and the load. A seesaw is a classic example, where the plank pivots on a central fulcrum. When using a spoon to scoop food, one can think of the handle as the effort arm, the bowl of the spoon as the load arm, and the point where the handle meets the hand as the fulcrum. When you press down on the handle, the spoon pivots at the fulcrum (your hand), lifting the bowl and the food within it. This configuration allows you to apply a small force at the handle to lift a larger load at the spoon's bowl.

Second-Class Lever:

In a second-class lever, the load is situated between the effort and the fulcrum. A wheelbarrow exemplifies this type, where the wheel acts as the fulcrum, the load is in the wheelbarrow, and the effort is applied at the handles. Using a spoon to crack open a soft-boiled egg can illustrate this principle. If you place the bowl of the spoon on the egg (the load) and push down on the handle (the effort), with the edge of the egg cup acting as the fulcrum, you are effectively using a second-class lever. The force you apply at the handle is magnified at the bowl, making it easier to crack the egg open.

Third-Class Lever:

In a third-class lever, the effort is applied between the load and the fulcrum. This type is common in the human body, such as the action of the biceps muscle when lifting a weight. When stirring a pot of soup with a spoon, your hand holding the handle acts as the fulcrum, the point where your fingers grip the spoon applies the effort, and the resistance of the soup acts as the load. This configuration prioritizes speed and range of motion over force, allowing you to stir the contents quickly and effectively.

The design of a spoon also optimizes leverage. The handle's length and shape provide the necessary mechanical advantage for various tasks. A longer handle increases the lever arm, which reduces the effort needed to lift or stir a load. The ergonomic shape ensures that the spoon is comfortable to hold and use for extended periods without causing fatigue or discomfort.

Material choice is another crucial aspect. Spoons are made from a variety of materials, including metal, wood, and plastic, each with specific properties that influence the utensil's leverage efficiency. Metal spoons, for instance, are rigid and durable, providing a sturdy fulcrum and lever arm. Wooden spoons are lighter and provide a good grip, making them ideal for stirring and mixing. Plastic spoons are often used for disposable purposes but still provide sufficient leverage for eating and simple tasks.

The versatility of a spoon as a lever extends to its use in culinary arts. Chefs and home cooks alike use spoons not only for eating but also for measuring, mixing, and serving. Measuring spoons utilize the same principles of leverage to ensure precise amounts of ingredients are used. The consistent shape and size of the spoon's bowl provide a standard measure, while the handle offers leverage for scooping and leveling.

In mixing and stirring, the spoon's leverage allows for efficient motion within the mixing bowl or pot. The handle acts as the lever arm, with the user's hand providing the fulcrum. The effort applied at the handle translates into effective mixing action at the spoon's bowl. This process is essential for evenly distributing ingredients and achieving the desired consistency in culinary preparations.

The spoon's role in serving food also exemplifies leverage. When transferring food from a pot to a plate, the spoon acts as a lever to lift and move the load. The design of serving spoons often includes a broader bowl and a longer handle to maximize the mechanical advantage, allowing for larger quantities to be served with minimal effort.

Beyond the kitchen, the principles of leverage applied by a spoon can be seen in various other contexts. For example, in medical settings, spoons are used to administer medicine, especially in pediatric care. The gentle curve and leverage provided by the spoon ensure accurate and safe delivery of liquid medications to patients.

In the context of physics education, the everyday spoon serves as an excellent tool to demonstrate and teach the principles of leverage. Students can conduct simple experiments using spoons to understand the different classes of levers and the concept of mechanical advantage.

These hands-on activities make abstract concepts tangible and relatable, fostering a deeper understanding of fundamental physics principles.

The evolution of spoon design over time also reflects an understanding of leverage and ergonomics. Ancient spoons were often simple scoops made from natural materials like shells or wood. As civilizations advanced, so did the design and materials of spoons, with metals like bronze, silver, and stainless steel becoming common. Modern spoons incorporate ergonomic designs that optimize leverage, ensuring they are comfortable to use while maximizing efficiency.

In summary, the everyday spoon is a marvel of leverage principles at work. It embodies the laws of physics through its simple yet effective design, demonstrating the power of levers in our daily lives. From eating and cooking to educational tools and medical applications, the spoon's ability to amplify force with minimal effort showcases the elegance of leverage. By understanding the science behind this common utensil, we gain a greater appreciation for the intricate balance of form and function that makes our daily tasks easier and more efficient.

Chapter 4: Tapping into Touchscreens: Capacitive Technology

Touchscreen technology has revolutionized how we interact with electronic devices, becoming a ubiquitous part of modern life. Whether on smartphones, tablets, ATMs, or even in-car navigation systems, touchscreens have transformed the user interface, making it more intuitive and accessible. Among the various types of touchscreen technologies, capacitive technology stands out for its precision, responsiveness, and widespread adoption.

Capacitive touchscreen technology is based on the principle of capacitance, which is the ability of a system to store an electric charge. Capacitance is measured in farads and is determined by the characteristics of the materials and the area over which the charge is distributed. In the context of a touchscreen, capacitive sensing relies on detecting changes in the electrical properties of the screen when a conductive object, such as a human finger, comes into contact with it.

At the core of a capacitive touchscreen is a layer of conductive material, typically made of indium tin oxide (ITO), which is applied as a thin, transparent film to the underside of the glass screen. This conductive layer is arranged in a grid pattern, with horizontal and vertical lines that create a matrix of capacitors. When the screen is powered, an electric field is established across this grid. The touchscreen controller continuously measures the capacitance at each point in the grid.

When a user touches the screen, their finger, being conductive, interacts with the electric field. This interaction causes a local change in capacitance at the point of contact. The touchscreen controller detects this change by monitoring the variations in the electric field. Because the grid pattern allows for precise localization of the change in capacitance, the controller can determine the exact coordinates of the touch. This data is then processed by the device's operating system to execute the corresponding action, such as opening an app or typing a character.

There are two primary types of capacitive touchscreens: surface capacitive and projected capacitive. Surface capacitive touchscreens are simpler and were among the first to be developed. In these screens, a single conductive layer is used, and the system detects changes in capacitance on the surface of the screen. However, surface capacitive screens are less sensitive and can only register a single touch at a time.

Projected capacitive touchscreens, on the other hand, are more advanced and are the standard in modern devices. These screens use multiple layers of conductors arranged in a grid pattern, which allows for multi-touch capability. Multi-touch technology enables the screen to detect and respond to multiple points of contact simultaneously. This feature is essential for gestures such as pinch-to-zoom, swipe, and multi-finger typing, which have become integral to the user experience on smartphones and tablets.

The construction of a projected capacitive touchscreen involves several layers. Starting from the bottom, there is a glass substrate that provides structural support. On top of this substrate, a layer of ITO is deposited, forming the first set of conductive lines. A thin insulating layer separates this from a second layer of ITO, which forms the second set of conductive lines, oriented perpendicular to the first set. This grid of conductors is then covered by another layer of protective glass or plastic.

To ensure high sensitivity and accuracy, advanced manufacturing techniques are employed to create these layers with precise alignment and minimal thickness. The transparency of the ITO layers is crucial because it ensures that the display underneath the touchscreen remains visible and clear. Additionally, the protective top layer must be durable enough to withstand repeated touches and swipes without scratching or degrading.

One of the significant advantages of capacitive touchscreens over other types, such as resistive touchscreens, is their superior touch sensitivity and accuracy. Resistive touchscreens rely on pressure applied to the screen to register a touch, which can make them less responsive and less accurate, especially for multi-touch gestures. Capacitive touchscreens, by detecting the electrical properties of the touch, can provide a more seamless and responsive user experience. This responsiveness is particularly important for applications that require quick and precise input, such as gaming or drawing.

Another advantage is the durability and ease of maintenance of capacitive touchscreens. Since they are typically made with glass surfaces, they are resistant to scratches and can be cleaned easily without damaging the touch functionality. This makes them ideal for use in environments where hygiene is critical, such as in medical devices or public kiosks.

Capacitive touchscreens also offer aesthetic and design benefits. The ability to detect touch through a solid surface allows for sleek, seamless designs without the need for physical buttons. This has led to the minimalist design trends seen in many modern electronic devices, where the entire front surface is a smooth, uninterrupted touchscreen.

Despite their many advantages, capacitive touchscreens also have some limitations. One of the primary challenges is their sensitivity to conductive materials, which can lead to unintended touches or the inability to register touches if the user is wearing gloves. However, advancements in capacitive technology have addressed some of these issues. For instance, special gloves designed for touchscreen use and improvements in touch controllers that can detect touches through gloves or other materials have mitigated these problems.

The capacitive touchscreen market continues to evolve with ongoing research and development aimed at improving performance and expanding capabilities. Innovations such as pressure-sensitive touchscreens, which can detect varying levels of force, add a new dimension to user interaction, enabling more nuanced inputs and controls. Flexible and foldable displays, which incorporate capacitive touch technology, are also emerging, promising new form factors and applications.

Chapter 5: Light Bulbs and Luminance: Shedding Light on Illumination

Light bulbs are an essential part of modern life, providing the illumination necessary for countless activities, from reading and working to enhancing the ambiance of our homes. The science behind light bulbs and luminance involves a deep understanding of physics, chemistry, and engineering. By exploring the different types of light bulbs, how they produce light, and the principles of luminance, we can appreciate the intricate processes that enable us to "shed light" on our surroundings.

The journey of artificial light began with the invention of the incandescent light bulb by Thomas Edison in the late 19th century. Incandescent bulbs work by passing an electric current through a thin tungsten filament. As the filament heats up to a high temperature (around 2,500 to 3,000 degrees Celsius), it emits visible light through a process known as incandescence. Incandescence is the emission of light from a hot object due to its temperature. The filament's resistance to the electric current causes it to heat up and glow, producing light.

One of the primary characteristics of incandescent bulbs is their simplicity and ability to produce a warm, continuous spectrum of light that closely resembles natural sunlight. However, they are not very energy efficient. A significant portion of the electrical energy is converted into heat rather than visible light, with only about 10% of the energy being used for illumination. This inefficiency led to the development of alternative lighting technologies that provide better energy utilization.

Fluorescent light bulbs emerged as an energy-efficient alternative to incandescent bulbs. These bulbs work on a different principle called fluorescence. Inside a fluorescent bulb, a low-pressure mercury vapor produces ultraviolet (UV) light when excited by an electric current. This UV light is not visible to the human eye but strikes the phosphor coating on the inside of the bulb. The phosphor absorbs the UV radiation and re-emits it as visible light. This process is known as phosphorescence. Fluorescent bulbs are much more efficient than incandescent bulbs, converting about 20-30% of the electrical energy into visible light.

Compact fluorescent lamps (CFLs) are a popular type of fluorescent bulb that offers the same energy efficiency in a more compact form factor. CFLs are designed to fit into standard light sockets, making them an easy replacement for incandescent bulbs. Despite their efficiency, fluorescent bulbs have some drawbacks, such as the presence of mercury, which poses environmental and health risks if the bulbs break or are improperly disposed of. Additionally, they can take a moment to reach full brightness and may have a cooler light spectrum that some people find less pleasing than the warm light of incandescent bulbs.

Light-emitting diodes (LEDs) represent the latest advancement in lighting technology. LEDs produce light through electroluminescence, a process where an electrical current passes through a semiconductor material, causing it to emit photons. The efficiency of LEDs is significantly higher than that of incandescent and fluorescent bulbs, with some LEDs converting up to 90% of the electrical energy into visible light. This high efficiency, combined with a long lifespan, makes LEDs the preferred choice for many lighting applications today.

LEDs are composed of multiple layers of semiconductor materials, typically including a positively charged (p-type) layer and a negatively charged (n-type) layer. When voltage is applied, electrons from the n-type layer and holes from the p-type layer move towards the junction between these layers. When an electron meets a hole, it falls into a lower energy state, releasing energy in the form of a photon. The wavelength (and thus the color) of the emitted light depends on the materials used in the semiconductor and their energy band gap.

One of the significant advantages of LEDs is their versatility in producing light of different colors and intensities. By combining multiple LEDs with different wavelengths, manufacturers can create light sources that emit white light or a wide range of colors. This capability is used in various applications, from decorative lighting to advanced display technologies.

Luminance, the measure of the amount of light emitted from a surface per unit area in a given direction, is a crucial concept in understanding how light bulbs illuminate our surroundings. It is measured in candelas per square meter (cd/m^2). Luminance differs from illuminance, which measures the amount of light incident on a surface, measured in lux (lx). While illuminance concerns how much light falls on a surface, luminance is about the light that is reflected or emitted from the surface.

Several factors affect the luminance of a light source, including its intensity, the distribution of light, and the reflectivity of surfaces in the environment. For instance, a highly reflective surface will have higher luminance than a dull, non-reflective surface under the same lighting conditions. Understanding luminance is essential in various fields, such as lighting design, photography, and vision science, as it directly impacts how we perceive brightness and contrast.

Color temperature is another important aspect of lighting, affecting both the functionality and ambiance of a space. Measured in kelvins (K), color temperature describes the appearance of the light emitted by a source. Lower color temperatures (around 2,700-3,000 K) produce a warm, yellowish light, similar to incandescent bulbs, which is often preferred for residential settings due to its cozy and relaxing effect.

Higher color temperatures (around 5,000-6,500 K) emit a cool, bluish light, resembling daylight, which is suitable for work environments and areas where clear visibility and focus are required.

The development of smart lighting systems has added another dimension to how we use light. These systems incorporate LEDs with advanced controls, allowing users to adjust the brightness, color temperature, and even the color of the light through smartphone apps or voice commands. Smart lighting enhances convenience, energy efficiency, and personalization, enabling users to create different lighting scenes for various activities and moods.

Lighting design is a field that blends art and science to optimize the use of light in spaces. Effective lighting design considers factors such as the purpose of the space, the tasks being performed, and the psychological and physiological effects of light on occupants. For example, in an office setting, good lighting design aims to reduce glare and shadows, provide adequate task lighting, and create a pleasant environment that promotes productivity and well-being. In residential settings, the focus might be on creating warm, inviting spaces that enhance relaxation and comfort.

Energy efficiency is a significant consideration in modern lighting design, driven by both environmental concerns and cost savings. Transitioning from incandescent bulbs to more efficient options like LEDs and CFLs can substantially reduce energy consumption and greenhouse gas emissions. Innovations in lighting technology continue to push the boundaries of efficiency and sustainability, such as the development of organic LEDs (OLEDs) and advanced lighting control systems that optimize energy use based on occupancy and natural light availability.

The environmental impact of lighting extends beyond energy consumption. The production, use, and disposal of light bulbs also have ecological implications. For instance, the use of hazardous materials like mercury in fluorescent bulbs necessitates careful handling and disposal to prevent environmental contamination. On the other hand, the long lifespan of LEDs reduces the frequency of replacement, thereby reducing waste. Recycling programs and advancements in manufacturing processes aim to minimize the environmental footprint of lighting technologies.

Chapter 6: The Dynamics of Doors: Hinges and Handles

The dynamics of doors, encompassing the intricate workings of hinges and handles, delve into a fascinating intersection of physics, engineering, and design. Doors are ubiquitous in our daily lives, providing security, privacy, and access control, but their seemingly simple operation relies on complex mechanical principles. Understanding the dynamics of doors involves exploring the types of hinges and handles, the forces and motion involved, and the materials and design considerations that ensure functionality and durability.

At the heart of a door's operation are the hinges, which allow it to swing open and closed. Hinges are simple mechanical bearings that connect two solid objects, allowing a limited range of rotational movement around a fixed axis. The basic principle of a hinge involves a pivot point, where one part (the leaf) is attached to the door and the other part to the door frame. The pivot allows the door to rotate, enabling it to open and close smoothly.

There are several types of hinges, each suited to specific applications and door designs. The most common type is the butt hinge, typically used in residential doors. Butt hinges consist of two rectangular metal plates (leaves) joined by a pin. One leaf is mortised into the door edge and the other into the door frame, allowing the door to pivot on the pin. The size and thickness of the hinge depend on the door's weight and intended use.

Continuous hinges, also known as piano hinges, run the entire length of the door, providing uniform support and reducing the stress on the door and frame. They are ideal for heavy or high-traffic doors, as they distribute the load evenly and minimize wear. Continuous hinges are commonly used in commercial and industrial settings.

Concealed hinges, often found in modern cabinetry and some doors, are designed to be hidden from view when the door is closed. These hinges provide a clean, uninterrupted appearance and are adjustable, allowing for precise alignment and smooth operation. Concealed hinges use a complex mechanism to provide full motion within a compact form factor, demonstrating advanced engineering in hinge design.

Pivot hinges are another type, typically used for heavy or wide doors that need to swing in both directions. These hinges are mounted at the top and bottom of the door rather than on the side, allowing the door to pivot around a central axis. Pivot hinges provide a smooth, elegant operation and are often seen in upscale homes and commercial buildings.

The forces involved in hinge operation are primarily related to the weight of the door and the friction between the hinge components. The hinge must support the door's weight while allowing it to move freely. This requires careful selection of materials and design to minimize friction and wear. Most hinges are made from durable metals like steel, brass, or stainless steel, chosen for their strength, corrosion resistance, and longevity.

In addition to hinges, door handles play a crucial role in the dynamics of doors. Handles provide the means to operate the door, offering both functional and aesthetic value. There are several types of door handles, each with specific design considerations and applications.

Lever handles are one of the most common types, especially in commercial and residential settings. A lever handle operates by pressing down or lifting up the lever, which retracts the latch mechanism inside the door, allowing it to open. Lever handles are favored for their ease of use, particularly for individuals with limited hand strength or dexterity.

They come in various styles and finishes, making them versatile for different architectural designs.

Knob handles are another popular option, often used in residential doors. Knob handles require a twisting motion to retract the latch. While they offer a classic look, they can be more challenging to operate for people with disabilities or young children. Knob handles are available in a wide range of designs and materials, from traditional brass to modern glass or ceramic.

Pull handles are typically found on sliding doors, cabinets, and commercial entrances. These handles do not operate a latch mechanism directly but provide a grip for pulling or sliding the door open. Pull handles can be simple bars or more elaborate designs, depending on the door's style and function.

Push plates and panic bars are specialized types of handles used primarily in public buildings for emergency exits. Push plates are flat panels that, when pushed, release the door latch. Panic bars, also known as crash bars, are horizontal bars installed across the door that, when pressed, quickly and easily open the door. These handles are designed for safety, allowing for quick evacuation in emergencies.

The dynamics of door handles involve both mechanical and ergonomic considerations. Mechanically, the handle must provide sufficient leverage to operate the latch or lock mechanism with minimal effort. This requires precise engineering to ensure smooth operation and durability. Ergonomically, the handle must be comfortable to use, considering factors such as grip size, shape, and placement.

Materials play a significant role in the design and functionality of door handles. Metals like brass, stainless steel, and aluminum are commonly used for their durability and aesthetic appeal. Brass, for example, is not only strong but also has antimicrobial properties, making it ideal for public and healthcare settings. Stainless steel is valued for its corrosion resistance and modern look. In some cases, materials like wood, glass, or ceramic are used for decorative handles, adding a unique touch to the door's design.

The integration of hinges and handles into the door system requires careful consideration of the door's overall design and use. For example, the placement and type of hinge can affect the door's swing and clearance, while the handle's design and position influence ease of use and accessibility. Proper installation is crucial to ensure the door operates smoothly and securely, with adequate support and alignment.

Advancements in technology have also influenced the dynamics of doors, particularly with the introduction of smart locks and automated systems. Smart locks can be integrated into traditional door handles, allowing for keyless entry using smartphones, keypads, or biometric sensors. These systems enhance security and convenience, offering features like remote access, activity logs, and customizable access codes.

Automated doors, commonly used in commercial and public buildings, utilize sensors and motors to open and close the doors automatically. These systems often incorporate motion detectors, pressure sensors, or touchless activation to enhance accessibility and hygiene. Automated doors require sophisticated control systems and robust mechanical components to ensure reliable and safe operation.

The principles of physics are deeply embedded in the dynamics of doors, particularly in understanding the forces and motion involved. When a door swings open or closed, it acts as a lever, with the hinges serving as the pivot point. The force applied to the handle is transferred through the door to the hinges, causing it to rotate around the pivot. The amount of force needed to move the door depends on its weight, the distance from the hinges (lever arm length), and the friction in the hinges.

Friction is a key factor in the operation of both hinges and handles. In hinges, friction must be minimized to allow smooth movement and reduce wear. This is often achieved through the use of lubricants, ball bearings, or low-friction materials like nylon or Teflon in the hinge design. In door handles, friction can provide tactile feedback that helps users sense the operation of the latch mechanism.

The dynamics of doors also encompass the acoustic and thermal properties of door systems. Doors can serve as barriers to sound and heat, depending on their construction and materials. Solid-core doors, for example, provide better sound insulation than hollow-core doors, making them suitable for bedrooms or offices. Thermal insulation can be enhanced with weatherstripping, insulated cores, or materials like fiberglass.

Chapter 7: The Mechanics of Faucets: Fluid Dynamics at Work

The mechanics of faucets, an essential fixture in homes and public spaces, offer a fascinating insight into fluid dynamics, mechanical engineering, and material science. Faucets are designed to control the flow of water from a plumbing system, providing a regulated stream for various uses, from washing hands and dishes to filling containers. Understanding how faucets work requires an in-depth exploration of their internal mechanisms, the principles of fluid dynamics they employ, and the advancements in technology that have enhanced their efficiency and functionality.

At its core, a faucet is a valve that controls the release of water. The primary function of a faucet is to regulate water flow and temperature through the manipulation of its internal components. The essential parts of a faucet include the spout, the handle(s), the valve, the aerator, and the supply lines. The design and operation of these components can vary significantly depending on the type and purpose of the faucet.

Faucets can be broadly categorized into two types based on their operation: compression faucets and non-compression faucets. Compression faucets are the oldest type, relying on rubber washers to create a seal that stops water flow. When the handle is turned, the washer is compressed against a valve seat, blocking the passage of water. Turning the handle in the opposite direction lifts the washer off the seat, allowing water to flow. While simple and effective, compression faucets can be prone to wear and require regular maintenance to replace worn-out washers.

Non-compression faucets, which include ball, cartridge, and ceramic disk faucets, offer more advanced mechanisms for controlling water flow. Ball faucets, commonly found in kitchens, use a single handle that moves over a ball-shaped cap. The movement of the handle changes the position of the ball, which has various slots and holes that align with the inlet ports to control the mix of hot and cold water and regulate the flow. The design allows for smooth operation and easy adjustment of water temperature and flow rate.

Cartridge faucets, available in single-handle and double-handle designs, use a cylindrical cartridge to control water flow. The cartridge moves up and down to regulate flow and rotates to adjust the temperature. Single-handle cartridge faucets are popular for their ease of use and precise control, while double-handle models offer separate controls for hot and cold water. Cartridge faucets are known for their reliability and ease of repair, as the cartridge can be easily replaced if it becomes worn or damaged.

Ceramic disk faucets represent the latest advancement in faucet technology, providing durability and smooth operation. These faucets feature two ceramic disks with precisely machined surfaces that slide over each other to control water flow. The disks have holes that align to allow water through when the handle is moved. The hard ceramic material is highly resistant to wear and corrosion, making ceramic disk faucets one of the most reliable and long-lasting options available.

The operation of faucets is deeply rooted in the principles of fluid dynamics, which study the behavior of fluids (liquids and gases) in motion. One of the key principles at play in faucet mechanics is Bernoulli's principle, which states that an increase in the velocity of a fluid results in a decrease in pressure. This principle is fundamental in understanding how faucets regulate water flow and pressure.

When water flows through a faucet, it is subject to various changes in pressure and velocity. The design of the faucet's internal passages and the valve mechanism creates a controlled flow that can be adjusted by the user. The shape and size of the spout, as well as the inclusion of features like aerators, significantly influence the flow characteristics.

Aerators are small mesh screens or perforated disks attached to the end of the faucet spout. They play a crucial role in improving water flow efficiency and user experience. By mixing air with the water stream, aerators reduce the overall water volume while maintaining a strong and even flow. This not only conserves water but also minimizes splashing and creates a softer, more pleasant stream. Aerators can also act as filters, trapping debris and preventing it from flowing out of the spout.

Another important aspect of fluid dynamics in faucet design is laminar flow versus turbulent flow. Laminar flow occurs when water flows in parallel layers with minimal disruption between them, resulting in a smooth and steady stream. Turbulent flow, on the other hand, involves chaotic and irregular motion, leading to a more aerated and often noisier stream. Faucets are designed to optimize the flow type for their intended use, with many aiming to achieve a balance between laminar and turbulent flow for efficiency and comfort.

The materials used in faucet construction are critical to their performance and longevity. Common materials include brass, stainless steel, and various plastics and composites. Brass is widely used for its durability, resistance to corrosion, and ability to be easily machined and polished. Stainless steel offers excellent corrosion resistance and a modern aesthetic, making it popular in contemporary designs. Advanced plastics and composites are used in components that require precision molding and resistance to wear and chemical exposure.

In addition to mechanical components and fluid dynamics, the evolution of faucets has seen significant advancements in technology, particularly with the advent of smart faucets and touchless designs. Smart faucets incorporate electronic sensors and microprocessors to provide enhanced functionality and convenience. These faucets can be activated by touch or motion sensors, allowing for hands-free operation. This not only improves hygiene by reducing contact with potentially contaminated surfaces but also offers greater water conservation by automatically shutting off the flow when not in use.

Touchless faucets use infrared sensors to detect the presence of hands or objects beneath the spout. When the sensor is triggered, an electronic valve opens to allow water flow and closes when the sensor is no longer activated. This technology is commonly used in public restrooms, kitchens, and healthcare settings, where hygiene and water conservation are paramount.

Another innovation in faucet technology is the integration of water filtration systems. Some modern faucets come equipped with built-in filters that remove impurities and contaminants from the water supply, providing clean and safe drinking water directly from the tap. These filtration systems can include activated carbon filters, reverse osmosis membranes, and other advanced filtration media that improve water quality and taste.

The design of faucets also takes into account ergonomic considerations, ensuring that they are easy and comfortable to use. Handles and levers are designed to be operated with minimal effort, accommodating users of all ages and abilities. The height and reach of the spout are carefully considered to provide adequate clearance for various tasks, such as filling pots or washing hands, without causing excessive splashing.

Chapter 8: The Wonder of Windows: Glass and Thermal Insulation

Windows, often described as the eyes of a building, serve multiple crucial functions beyond their aesthetic appeal. They allow natural light to enter, provide views of the outside world, and play a significant role in the thermal performance and energy efficiency of a structure. Understanding the intricacies of windows involves delving into the properties of glass, the principles of thermal insulation, and the technological advancements that enhance their performance. This comprehensive exploration reveals the wonder of windows and their impact on our comfort, energy consumption, and environmental footprint.

The fundamental component of a window is its glass. Glass, a unique material with both solid and liquid properties, is primarily composed of silica (sand), soda ash, and lime, which are melted together at high temperatures to form a transparent and rigid structure. The production of glass involves a careful balance of these ingredients to achieve the desired properties, including clarity, strength, and thermal performance.

There are several types of glass used in windows, each designed to meet specific needs and conditions. Float glass, the most common type, is produced by floating molten glass on a bed of molten tin, creating a flat and uniform sheet. This method ensures high optical quality and consistency in thickness. Float glass can be further processed to enhance its properties, such as being tempered, laminated, or coated with low-emissivity (Low-E) coatings.

Tempered glass, also known as toughened glass, undergoes a heat treatment process that increases its strength and safety. The glass is heated to high temperatures and then rapidly cooled, creating internal stresses that make it much stronger than regular glass. When broken, tempered glass shatters into small, blunt pieces rather than sharp shards, reducing the risk of injury. This makes it ideal for applications where safety is a concern, such as in doors, windows, and skylights.

Laminated glass consists of two or more layers of glass bonded together with an interlayer, usually made of polyvinyl butyral (PVB) or ethylene-vinyl acetate (EVA). This construction enhances the glass's safety and security features. When impacted, the interlayer holds the broken pieces together, preventing them from scattering. Laminated glass also provides improved acoustic insulation and UV protection, making it a popular choice for windows in areas with high noise levels or strong sunlight.

Low-emissivity (Low-E) glass is specially coated to improve its thermal performance. The Low-E coating is a thin, invisible layer of metallic oxides applied to the glass surface. This coating reflects infrared radiation, keeping heat inside during the winter and outside during the summer. By reducing the amount of heat transfer through the window, Low-E glass helps maintain a consistent indoor temperature, enhancing comfort and energy efficiency.

Double-glazed and triple-glazed windows further enhance thermal insulation by incorporating multiple layers of glass with air or gas-filled spaces between them. These spaces act as insulators, reducing heat transfer and improving the window's overall thermal performance. Double-glazed windows have two panes of glass separated by a spacer, while triple-glazed windows have three panes and two spacers. The gaps between the panes can be filled with inert gases like argon or krypton, which have lower thermal conductivity than air, providing better insulation.

The spacers used in double-glazed and triple-glazed windows play a crucial role in their performance. Traditional spacers, made of aluminum or other metals, can conduct heat, creating thermal bridges that reduce the window's insulating properties. To address this issue, warm-edge spacers made from materials with lower thermal conductivity, such as plastic or stainless steel, are used. These spacers help minimize heat transfer, improving the window's overall energy efficiency.

Thermal insulation in windows is essential for maintaining indoor comfort and reducing energy consumption. Heat transfer through windows occurs via conduction, convection, and radiation. Conduction is the transfer of heat through the glass and frame materials. Convection occurs when air within the window assembly or room moves and carries heat with it. Radiation involves the transfer of heat in the form of infrared energy, which can pass through the glass.

To maximize thermal insulation, modern windows are designed to minimize all three modes of heat transfer. The use of Low-E coatings, multiple glazing layers, and inert gas fills reduces conduction and radiation. Proper sealing and weatherstripping around the window frame prevent air leakage, reducing convection. Additionally, advanced framing materials, such as insulated vinyl or fiberglass, offer better thermal performance than traditional wood or aluminum frames.

The installation of energy-efficient windows has a significant impact on a building's energy consumption and carbon footprint. According to the U.S. Department of Energy, heat gain and loss through windows account for 25-30% of residential heating and cooling energy use. By upgrading to energy-efficient windows, homeowners can reduce their energy bills and contribute to environmental sustainability. The benefits include improved indoor comfort, reduced noise levels, and protection from harmful UV rays, which can fade furniture and flooring.

Beyond thermal insulation, windows play a vital role in natural lighting and ventilation. Properly designed and placed windows maximize the use of natural light, reducing the need for artificial lighting and associated energy consumption. Daylighting, the practice of using natural light to illuminate interior spaces, has been shown to improve occupants' well-being, productivity, and mood. Large, strategically placed windows and the use of light shelves or reflective surfaces can enhance the distribution of natural light within a building.

Ventilation through windows is another critical aspect of indoor air quality and comfort. Operable windows allow for the exchange of indoor and outdoor air, providing fresh air and removing stale air, pollutants, and moisture. This natural ventilation helps regulate indoor temperatures and humidity levels, contributing to a healthier living environment. Advanced window designs, such as tilt-and-turn or casement windows, offer flexible ventilation options while maintaining security and energy efficiency.

Technological advancements have also led to the development of smart windows, which can dynamically control light and heat transmission. Electrochromic windows, for example, use a thin layer of electrochromic material that changes its opacity in response to an applied voltage. This allows the window to switch between clear and tinted states, reducing glare and heat gain while maintaining visibility. Smart windows can be integrated with building automation systems, enabling automated control based on occupancy, weather conditions, and user preferences.

In addition to electrochromic technology, other types of smart windows include photochromic and thermochromic windows. Photochromic windows darken in response to sunlight, similar to transition lenses in eyeglasses. Thermochromic windows change their tint based on temperature, providing passive control over heat gain and loss. These technologies offer the potential for significant energy savings and enhanced comfort by adapting to changing environmental conditions.

The environmental impact of windows extends beyond energy consumption. The materials and manufacturing processes used in window production can contribute to their overall sustainability. For instance, the extraction and processing of raw materials, such as silica for glass and metals for frames, can have significant environmental consequences. To address this, manufacturers are increasingly adopting sustainable practices, such as using recycled materials, reducing waste, and implementing energy-efficient production techniques.

Recycling and end-of-life disposal of windows also play a crucial role in their environmental footprint. Many components of windows, including glass, metal, and plastic, can be recycled, reducing the demand for virgin materials and minimizing waste. Innovative design approaches, such as modular and easily disassemblable windows, facilitate recycling and reuse. By considering the entire lifecycle of windows, from production to disposal, manufacturers and consumers can make more sustainable choices that benefit both the environment and society.

Chapter 9: Refrigerators and Heat Pumps: Keeping It Cool

Refrigerators and heat pumps are marvels of modern engineering, playing a crucial role in our daily lives by preserving food, maintaining comfort, and enhancing energy efficiency. Both appliances operate on the same fundamental principle of thermodynamics—transferring heat from one place to another—but they serve different purposes. Understanding how these devices work involves exploring the principles of refrigeration, the components and mechanics of each system, and the advancements that have improved their efficiency and functionality over time.

At the heart of both refrigerators and heat pumps is the refrigeration cycle, a process that relies on the physical properties of refrigerants to move heat. The refrigeration cycle consists of four main stages: evaporation, compression, condensation, and expansion. In a refrigerator, this cycle removes heat from the interior and releases it outside, while in a heat pump, the cycle can be reversed to either heat or cool a space.

The refrigeration cycle begins with the evaporator, where the refrigerant absorbs heat from the environment and evaporates into a low-pressure gas. Inside a refrigerator, the evaporator is located within the insulated compartment, absorbing heat from the food and air inside. The refrigerant, now a vapor, carries this heat to the compressor.

The compressor, often considered the heart of the refrigeration system, is responsible for increasing the pressure and temperature of the refrigerant gas. By compressing the refrigerant, the compressor raises its temperature above the ambient temperature outside the refrigerator or heat pump. This high-pressure, high-temperature gas then flows to the condenser.

In the condenser, the refrigerant releases the absorbed heat to the surroundings and condenses back into a liquid. For a refrigerator, this means releasing heat to the kitchen or the space behind the appliance. For a heat pump, the condenser can be located inside the building during heating mode, releasing warmth into the indoor environment. The refrigerant, now a high-pressure liquid, moves to the expansion valve.

The expansion valve, or expansion device, regulates the flow of refrigerant into the evaporator. As the high-pressure liquid passes through the valve, it expands and drops in pressure, causing its temperature to decrease. This cold, low-pressure liquid re-enters the evaporator, and the cycle begins anew.

Refrigerators are designed to keep the interior at a consistently low temperature to preserve perishable food items. Modern refrigerators use various technologies to enhance efficiency and functionality. For instance, many refrigerators feature variable-speed compressors that adjust their speed based on the cooling demand, reducing energy consumption. Additionally, advanced insulation materials and airtight door seals help maintain the internal temperature and prevent heat ingress.

Frost-free refrigerators, a significant advancement in refrigeration technology, use a defrost cycle to prevent ice buildup on the evaporator coils. This is achieved by periodically warming the coils to melt any accumulated frost, which then drains away. This not only improves efficiency by ensuring optimal heat transfer but also eliminates the need for manual defrosting, making the refrigerator more convenient to use.

The choice of refrigerant is crucial for the performance and environmental impact of refrigerators and heat pumps. Historically, chlorofluorocarbons (CFCs) were widely used due to their excellent thermodynamic properties. However, CFCs have been phased out due to their detrimental effect on the ozone layer. They were replaced by hydrochlorofluorocarbons (HCFCs) and eventually by hydrofluorocarbons (HFCs), which have no ozone-depleting potential but still contribute to global warming. Today, there is a growing emphasis on using low-global-warming-potential (GWP) refrigerants, such as hydrocarbons (e.g., propane, isobutane) and new synthetic alternatives, to reduce the environmental impact of refrigeration systems.

Heat pumps, which operate on the same refrigeration cycle principles, are versatile devices capable of providing both heating and cooling. During the cooling mode, a heat pump functions like an air conditioner, removing heat from the indoor environment and releasing it outside. In heating mode, the cycle is reversed, and the heat pump extracts heat from the outside air (even in cold weather) and releases it inside the building.

The ability of heat pumps to transfer heat rather than generate it makes them highly efficient for heating applications. Unlike conventional heating systems, such as furnaces or electric heaters that convert fuel or electricity directly into heat, heat pumps move heat from one place to another, using much less energy. This efficiency is often expressed as the coefficient of performance (COP), which measures the ratio of heating or cooling output to the energy input. A typical heat pump can achieve a COP of 3 to 4, meaning it delivers three to four units of heat for every unit of electricity consumed.

Advancements in heat pump technology have led to the development of air-source, ground-source (geothermal), and water-source heat pumps, each with its advantages and applications. Air-source heat pumps, the most common type, extract heat from the outdoor air. They are relatively easy to install and are suitable for a wide range of climates.

However, their efficiency decreases as the outdoor temperature drops, making them less effective in extremely cold conditions.

Ground-source heat pumps, also known as geothermal heat pumps, utilize the stable temperature of the ground to provide heating and cooling. They involve burying a network of pipes (ground loop) in the ground or submerging them in a body of water. The ground loop circulates a refrigerant or a water-antifreeze solution, which absorbs heat from the ground in winter and dissipates heat into the ground in summer. Ground-source heat pumps are highly efficient and reliable, with lower operating costs over time, but they require a higher initial investment and more space for installation.

Water-source heat pumps operate similarly to ground-source systems but use a body of water, such as a lake, river, or well, as the heat exchange medium. They are particularly effective in areas with abundant water resources and can offer high efficiency and performance.

In addition to residential applications, heat pumps are increasingly used in commercial and industrial settings for space heating, cooling, and water heating. Large-scale heat pump systems can provide district heating, supplying multiple buildings with efficient and sustainable thermal energy. Industrial heat pumps are used in processes requiring heating and cooling, such as food processing, chemical manufacturing, and waste heat recovery.

The integration of heat pumps with renewable energy sources, such as solar and wind power, further enhances their sustainability and reduces greenhouse gas emissions. Solar-assisted heat pumps, for instance, use solar collectors to preheat the refrigerant or water before it enters the heat pump, improving overall efficiency. Similarly, heat pumps can be powered by electricity generated from wind turbines, providing a clean and renewable source of energy for heating and cooling.

Another area of innovation in heat pump technology is the development of smart and connected systems. Smart heat pumps can be integrated with home automation and energy management systems, allowing users to control and monitor their operation remotely. These systems can optimize performance based on weather conditions, occupancy patterns, and energy prices, enhancing comfort and reducing energy costs.

Thermal storage is another technique used to improve the efficiency and flexibility of heat pump systems. By storing excess heat or cold energy in thermal storage units, heat pumps can operate more efficiently during off-peak periods and provide consistent heating or cooling even during high-demand times. Thermal storage can be in the form of hot water tanks, phase-change materials, or underground thermal energy storage systems.

In summary, refrigerators and heat pumps are essential appliances that rely on the principles of thermodynamics and refrigeration to provide cooling and heating. The refrigeration cycle, involving evaporation, compression, condensation, and expansion, is the core mechanism that enables these devices to transfer heat efficiently. Advancements in technology, such as variable-speed compressors, Low-GWP refrigerants, and smart controls, have significantly improved the performance and environmental sustainability of refrigerators and heat pumps. By leveraging renewable energy sources and integrating with advanced energy management systems, these devices continue to evolve, offering greater efficiency, comfort, and convenience in our daily lives. As we move towards a more sustainable future, the continued innovation and adoption of efficient refrigeration and heat pump technologies will play a crucial role in reducing energy consumption and mitigating climate change.

Chapter 10: Toasters and Thermal Conductivity

Toasters, a common kitchen appliance, are marvels of thermal engineering that leverage the principles of thermal conductivity to transform slices of bread into perfectly toasted treats. The process of toasting involves the intricate interaction of heat transfer mechanisms, material properties, and precise control systems. Delving deeply into how toasters work reveals a fascinating interplay of science and technology that ensures consistent, efficient, and safe operation.

At its core, a toaster operates by converting electrical energy into heat energy through resistive heating elements, typically made of nichrome wire. Nichrome, an alloy of nickel and chromium, is chosen for its high electrical resistance and excellent thermal conductivity. When an electric current passes through these wires, they heat up due to the resistance they offer to the flow of electricity. This heat is then transferred to the bread primarily through radiation and convection, leading to the toasting process.

The heating elements in a toaster are designed to produce infrared radiation, which penetrates the surface of the bread and causes the Maillard reaction. This chemical reaction between amino acids and reducing sugars gives toasted bread its characteristic flavor, color, and texture. The efficiency of this process depends significantly on the material properties and design of the heating elements. Nichrome is particularly effective because it can withstand high temperatures without oxidizing or degrading, ensuring longevity and consistent performance.

To understand the role of thermal conductivity in a toaster, it's essential to explore how heat is transferred from the heating elements to the bread. Thermal conductivity is a measure of a material's ability to conduct heat. In the context of a toaster, it is crucial for the heating elements to have high thermal conductivity to ensure that heat is evenly distributed across the surface of the bread. Nichrome's thermal conductivity allows it to rapidly transfer heat to the surrounding air, which then carries the heat to the bread through convection.

The design of the toaster's internal components also plays a vital role in optimizing heat transfer. The placement of the heating elements is crucial; they must be positioned close enough to the bread to maximize heat transfer but far enough to prevent burning or uneven toasting. Many toasters use reflective surfaces behind the heating elements to direct more radiant heat towards the bread, enhancing the efficiency of the toasting process. These reflective surfaces are often made of materials with high reflectivity and low thermal conductivity to ensure that the heat is directed towards the bread rather than being absorbed by the toaster's interior components.

Modern toasters incorporate various technologies to control and optimize the toasting process. One common feature is the use of a thermostat or an electronic timer to regulate the toasting duration. The thermostat measures the temperature inside the toaster and adjusts the heating elements' power to maintain a consistent temperature. This ensures that the bread is toasted to the desired level of doneness without burning. Some advanced toasters use microprocessors to provide more precise control over the toasting process, allowing users to select specific settings for different types of bread or desired toast levels.

The insulation of the toaster's body is another critical aspect that affects thermal conductivity and energy efficiency. The exterior of the toaster is typically made of materials with low thermal conductivity, such as plastic or stainless steel, to prevent heat from escaping and to keep the outer surface cool to the touch. This not only improves energy efficiency by directing more heat towards the bread but also enhances safety by reducing the risk of burns during operation.

Heat transfer within the toaster also involves convective currents, which are influenced by the design of the toaster's slots and ventilation. As the heating elements warm the air inside the toaster, this hot air rises and circulates around the bread, transferring heat through convection. To optimize this process, toasters are designed with carefully positioned slots and ventilation openings that promote efficient airflow. This ensures that the heat is evenly distributed around the bread, resulting in uniform toasting.

The choice of materials for the toaster's construction extends beyond the heating elements and exterior. The internal components, such as the crumb tray and the bread lifter mechanism, must also be made from materials that can withstand repeated heating cycles without degrading. Stainless steel and aluminum are commonly used for these parts due to their durability and thermal conductivity. These materials ensure that the toaster operates reliably over its lifespan and that heat is effectively managed within the appliance.

Safety features are integral to modern toaster design, leveraging principles of thermal conductivity to prevent overheating and fire hazards. Many toasters include thermal fuses or cut-off switches that interrupt the electrical circuit if the toaster exceeds a safe operating temperature. These components are made from materials that react predictably to changes in temperature, ensuring that the toaster shuts down in case of a malfunction. Additionally, the toaster's design often includes heat-resistant materials and insulation to contain heat within the appliance and prevent external surfaces from becoming dangerously hot.

Energy efficiency is another area where advancements in toaster technology and understanding of thermal conductivity have made a significant impact. Toasters are designed to heat up quickly and maintain consistent temperatures with minimal energy loss. This is achieved through the use of high-quality materials for the heating elements and insulation, as well as efficient design of the internal components to maximize heat transfer to the bread while minimizing energy consumption. Some toasters are equipped with energy-saving features, such as automatic shut-off and adjustable power settings, which further enhance their efficiency.

In addition to conventional pop-up toasters, toaster ovens offer more versatile cooking options by combining the principles of toasting with those of baking and broiling. Toaster ovens use larger heating elements and often include fans to circulate hot air, providing more uniform heat distribution. The design of toaster ovens must carefully balance thermal conductivity and insulation to ensure that they can perform a variety of cooking tasks efficiently. Advanced toaster ovens may include digital controls, programmable settings, and even convection cooking modes to provide precise and versatile cooking options.

The evolution of toaster design reflects ongoing advancements in material science and engineering, particularly in the realm of thermal conductivity. Researchers continue to explore new materials and coatings that can improve the efficiency and performance of toasters. For instance, the development of advanced ceramic coatings for heating elements could enhance their thermal conductivity and durability, leading to even more efficient and longer-lasting appliances.

Chapter 11: The Physics of Plugs: Electrical Connectivity

Plugs and electrical connectivity form the backbone of our modern, electrified world. They serve as the critical interface between electrical devices and the power sources that energize them. Understanding the physics behind plugs involves a detailed exploration of electrical principles, materials science, mechanical design, and safety features. This comprehensive examination covers the anatomy of plugs, the electrical principles governing their operation, the materials used in their construction, and the safety mechanisms that protect users and devices.

At its core, a plug is a device designed to establish a secure electrical connection between an appliance and an electrical outlet. The basic design of a plug includes prongs or pins that insert into corresponding slots in an outlet, allowing current to flow from the power source to the appliance. This seemingly simple action involves several complex physical and electrical principles that ensure safe and reliable connectivity.

Electrical plugs come in various designs and standards, differing by country and region. The most common types include Type A and B plugs in North America, Type C plugs in Europe, and Type G plugs in the United Kingdom. Despite these variations, all plugs share fundamental design principles aimed at ensuring compatibility with their respective outlets and providing safe electrical connections.

The primary function of a plug is to conduct electrical current from the outlet to the appliance. This process relies on the principles of electrical conductivity and Ohm's Law, which states that the current (I) flowing through a conductor is directly proportional to the voltage (V) across it and inversely proportional to the resistance (R) of the conductor

$(I = V/R)$. To minimize resistance and ensure efficient current flow, plugs and their internal components are typically made of materials with high electrical conductivity, such as copper or brass.

Copper is the preferred material for electrical contacts within plugs due to its excellent conductivity, ductility, and resistance to corrosion. Brass, an alloy of copper and zinc, is also commonly used because it combines good conductivity with greater mechanical strength and durability. These materials ensure that plugs can conduct electrical current efficiently while withstanding the mechanical stresses of repeated insertion and removal.

The mechanical design of plugs is crucial for ensuring a secure and reliable connection. The prongs or pins must fit snugly into the outlet slots to maintain firm contact and prevent arcing, which can occur if the connection is loose. Arcing is the phenomenon where an electric current flows through the air gap between conductors, creating a visible spark. This can cause overheating, damage to the plug and outlet, and potential fire hazards. To prevent arcing, plugs are designed with precise tolerances to ensure a tight fit and consistent contact pressure.

The shape and configuration of plug prongs are also designed to ensure proper orientation and polarity. For instance, Type A and B plugs have two flat prongs, with Type B also including a round grounding prong. The grounding prong is longer, ensuring it connects first when the plug is inserted, providing an immediate path to ground. This feature enhances safety by preventing electrical shock in case of a fault. The grounding prong also stabilizes the plug, reducing the risk of disconnection due to movement or vibration.

The insulating materials used in plug construction are equally important. The outer casing of a plug is typically made of plastic or rubber, materials chosen for their electrical insulating properties and resistance to heat. These materials prevent accidental contact with live electrical parts and protect the internal components from environmental factors such as moisture and dust. High-quality insulation is crucial for preventing short circuits and electrical shocks.

Inside the plug, the wires connecting the prongs to the appliance are also insulated to prevent accidental contact and ensure safe operation. These wires are typically made of stranded copper, which offers flexibility and durability. The insulation material, often PVC (polyvinyl chloride) or rubber, is selected for its resistance to heat, chemicals, and abrasion, ensuring the plug can withstand the rigors of everyday use.

Safety features in plugs are paramount, given the potential hazards associated with electricity. One key safety mechanism is the fuse, commonly found in plugs used in the United Kingdom and other regions with similar standards. A fuse is a safety device that protects electrical circuits by interrupting the flow of current if it exceeds a safe level. The fuse contains a thin wire that melts and breaks the circuit if the current becomes too high, preventing overheating and potential fire.

Another important safety feature is the ground connection, which provides a direct path for electrical current to return to the ground in case of a fault. This grounding protects users from electrical shock and helps stabilize the voltage within the system. Grounded plugs, such as Type B and Type G, include an additional prong connected to the ground wire, which is connected to the earth at the electrical panel. This ensures that any stray current is safely diverted away from the user and appliance.

Childproofing features are also incorporated into modern plug designs to enhance safety. Tamper-resistant outlets, for example, have built-in shutters that block access to the contacts unless a proper plug is inserted. This prevents children from inserting foreign objects into the outlet and receiving an electrical shock. Similarly, some plugs and outlets include built-in covers or caps that can be closed when not in use, providing an additional layer of protection.

Beyond the basic design and safety features, the physics of plugs also involves understanding the impact of electrical load and thermal effects. Electrical devices draw varying amounts of current, depending on their power requirements. High-power devices, such as heaters or air conditioners, require plugs and outlets capable of handling higher currents without overheating. Overloading a plug or outlet can cause excessive heat buildup due to the resistance in the conductors, potentially leading to melting, fire, or equipment damage.

To address these issues, plugs and outlets are rated for specific current and voltage levels. Common ratings include 10A, 13A, and 15A for residential plugs, indicating the maximum current they can safely handle. These ratings are based on rigorous testing and standards set by organizations such as the International Electrotechnical Commission (IEC) and Underwriters Laboratories (UL). Using plugs and outlets within their rated capacity ensures safe and reliable operation.

Thermal conductivity plays a significant role in the design of plugs, particularly in managing heat dissipation. Metals used in the electrical contacts, such as copper and brass, have high thermal conductivity, allowing them to efficiently dissipate heat generated by the flow of current. This prevents overheating and ensures that the plug remains cool to the touch during operation. The insulating materials used in plugs are also designed to withstand and dissipate heat, maintaining their integrity and preventing thermal degradation.

Advancements in materials science and engineering have led to the development of innovative plug designs and features that enhance safety, efficiency, and convenience. For instance, some modern plugs incorporate surge protection to safeguard electronic devices from voltage spikes caused by lightning strikes or power surges. Surge protectors use components such as metal oxide varistors (MOVs) to absorb and dissipate excess voltage, protecting sensitive electronics from damage.

Another innovation is the development of smart plugs, which integrate wireless communication technology to provide remote control and monitoring of electrical devices. Smart plugs connect to home automation systems via Wi-Fi or Bluetooth, allowing users to control appliances using a smartphone app or voice commands. These plugs often include energy monitoring features, providing real-time data on power consumption and helping users manage their energy usage more efficiently.

The evolution of plug design also reflects the growing emphasis on sustainability and energy efficiency. For example, eco-friendly plugs and outlets are designed to minimize standby power consumption, also known as "phantom load," which occurs when devices continue to draw power even when turned off. Advanced circuitry and design features in these plugs reduce or eliminate standby power usage, contributing to lower energy bills and reduced environmental impact.

In summary, the physics of plugs encompasses a wide range of principles and technologies that ensure safe and reliable electrical connectivity. From the choice of conductive materials and the design of prongs and insulation to the incorporation of safety features and advancements in smart technology, plugs are complex devices that play a crucial role in our daily lives. By understanding the underlying physics and engineering behind plugs, we can appreciate the sophisticated design and innovation that enable the seamless and safe operation of our electrical devices. As technology continues to evolve, the future of plugs promises even greater advancements in safety, efficiency, and functionality, further enhancing our ability to harness the power of electricity.

Chapter 12: Microwaves: Waves and Heating Mechanisms

Microwaves, an essential appliance in modern kitchens, revolutionized the way we cook and reheat food. Understanding how microwaves work involves delving into the physics of electromagnetic waves, the mechanisms of heat transfer, and the technological advancements that have made these appliances both efficient and safe. The science behind microwave ovens encompasses several key principles, including electromagnetic wave generation, wave-material interaction, and the resulting thermal processes that heat and cook food.

At the heart of a microwave oven is the magnetron, a device that generates microwaves, which are a form of electromagnetic radiation with frequencies typically around 2.45 gigahertz (GHz). The magnetron converts electrical energy into microwaves through the interaction of electrons and magnetic fields. When electrical current passes through the magnetron, it causes electrons to spiral within a magnetic field, producing high-frequency electromagnetic waves. These microwaves are then directed into the cooking chamber of the oven through a waveguide, which ensures that the waves are evenly distributed.

The microwave oven's ability to heat food quickly and efficiently is due to the unique properties of microwaves and their interaction with water molecules. Microwaves have wavelengths that are particularly effective at penetrating food and causing water molecules within the food to vibrate. Water molecules are polar, meaning they have a positive charge at one end and a negative charge at the other. When exposed to microwaves, these polar molecules attempt to align with the rapidly oscillating electric field of the waves. This constant reorientation generates friction and, consequently, heat through a process known as dielectric heating.

Dielectric heating is a form of heating that occurs when a dielectric material, such as water, absorbs electromagnetic energy and converts it into thermal energy. This mechanism is highly efficient for heating food because most foods contain a significant amount of water. As the microwaves penetrate the food, they cause water molecules to vibrate vigorously, producing heat that cooks the food from the inside out. This is in contrast to conventional ovens, which rely on conduction and convection to transfer heat from the outer surfaces to the interior of the food, often resulting in slower and less uniform cooking.

The design of a microwave oven includes several components that enhance its heating efficiency and ensure even cooking. The turntable, a rotating platform within the cooking chamber, helps distribute microwaves more uniformly across the food. By rotating the food, the turntable minimizes cold spots and ensures that all parts of the food are exposed to microwaves. Additionally, the walls of the microwave oven are lined with metal to reflect microwaves, preventing them from escaping and ensuring they are directed back into the food.

The microwave oven's control system regulates the power and duration of microwave generation to achieve the desired cooking results. Users can set the cooking time and power level, which adjusts the intensity and duration of the microwaves produced by the magnetron. Power levels are typically controlled by cycling the magnetron on and off, rather than varying the output power of the magnetron itself. This method, known as pulse-width modulation, allows for precise control over the cooking process and prevents overheating or overcooking of food.

Safety is a paramount consideration in microwave oven design. One of the key safety features is the interlock mechanism, which ensures that the microwave oven cannot operate when the door is open. This prevents exposure to microwaves, which can be harmful to humans if directly exposed. The door of the microwave oven is equipped with a metal mesh screen that allows users to see inside while effectively blocking microwaves from escaping. This screen works because the holes in the mesh are much smaller than the wavelength of the microwaves, thus reflecting the waves back into the cooking chamber.

The materials used in microwave-safe containers and cookware are specifically chosen for their ability to interact safely with microwaves. Microwave-safe materials, such as certain plastics, glass, and ceramics, do not absorb microwaves significantly and do not heat up directly. Instead, they allow microwaves to pass through them and heat the food inside. Conversely, materials like metal can reflect microwaves and cause arcing, which is a form of electrical discharge that can damage the microwave oven and pose a fire hazard. This is why it's essential to use only microwave-safe containers and avoid placing metal objects inside the oven.

Microwave ovens also incorporate thermal sensors and control algorithms to ensure that food is cooked evenly and safely. Advanced models may include moisture sensors that detect the steam emitted from food as it cooks, adjusting the power and cooking time accordingly. This feature helps prevent overcooking and ensures that food retains its moisture and texture. Some microwave ovens also feature inverter technology, which provides continuous and variable power control, allowing for more precise and even heating, especially for delicate foods.

The convenience and efficiency of microwave ovens have led to their widespread adoption, but they are also subject to continuous innovation and improvement. One area of development is the integration of microwave ovens with smart home technologies. Smart microwave ovens can be connected to the internet and controlled remotely via smartphone apps. These appliances often come with pre-programmed cooking settings and recipes, enabling users to cook a wide variety of dishes with minimal effort. Voice control and compatibility with virtual assistants, such as Amazon Alexa and Google Assistant, further enhance the convenience and user experience.

Another area of advancement is the development of combination microwave ovens, which integrate traditional cooking methods, such as convection and grilling, with microwave technology. These combination ovens offer greater versatility and cooking options, allowing users to bake, roast, and grill in addition to reheating and defrosting. By combining multiple cooking methods, these ovens can produce better culinary results, such as crispy exteriors and evenly cooked interiors, that are difficult to achieve with microwaves alone.

The environmental impact of microwave ovens is also an important consideration. Microwave ovens are generally more energy-efficient than conventional ovens for tasks such as reheating and cooking small portions of food. This efficiency is due to the direct interaction of microwaves with food, which reduces the time and energy required for cooking. However, the materials used in microwave oven construction, such as plastics and electronics, can pose environmental challenges if not properly managed at the end of the appliance's life cycle. Efforts to design more sustainable and recyclable microwave ovens are ongoing, with a focus on reducing waste and improving energy efficiency.

In summary, the physics of microwaves and their heating mechanisms encompass a wide range of scientific principles and technological innovations. The generation of microwaves by the magnetron, the interaction of microwaves with water molecules through dielectric heating, and the precise control of the cooking process all contribute to the efficiency and convenience of microwave ovens. Safety features, materials science, and continuous advancements in technology further enhance the performance and user experience of these ubiquitous kitchen appliances. As microwave ovens continue to evolve, they will likely incorporate even more sophisticated features and environmentally friendly designs, cementing their role as indispensable tools in modern culinary practices.

Chapter 13: From Ink to Paper: Printers and Electrophotography

Printers, particularly those that use electrophotography (laser printers), have become indispensable in both home and office environments. The process of transferring digital data onto physical paper involves a complex interplay of physics, chemistry, and engineering. Understanding the journey from ink or toner to paper requires a deep dive into the mechanisms and principles underlying electrophotographic printing technology.

Electrophotography, commonly known as laser printing, was invented by Chester Carlson in 1938 and has since evolved into a sophisticated technology used widely in modern printers. The process begins with the digital input, which could be a text document, image, or any other form of digital data. This data is processed by the printer's internal controller, converting it into a series of instructions that guide the printing process.

The core component of a laser printer is the photoreceptor drum, a cylindrical or belt-shaped object coated with a photosensitive material. This material is usually made of organic photoconductors (OPCs) or amorphous silicon, which have the property of changing their electrical conductivity when exposed to light. The drum starts out with a uniform negative charge, applied by a corona wire or a primary charge roller.

The laser unit plays a pivotal role in transferring the digital image onto the photoreceptor drum. The laser beam, controlled by a rotating polygonal mirror and lenses, scans across the surface of the drum in a precise manner. Wherever the laser beam hits, it discharges the photoconductive material, creating a latent electrostatic image. This image is a pattern of charges that corresponds to the digital data being printed.

The next step involves the application of toner, a fine powder made up of pigment and plastic particles. The toner is housed in a cartridge and is positively charged. As the toner particles pass by the photoreceptor drum, they adhere to the negatively charged areas that were exposed by the laser beam, forming a visible image. This step is known as development.

Once the toner image is formed on the photoreceptor drum, it must be transferred to paper. The transfer process involves a transfer corona wire or roller that applies a positive charge to the back of the paper as it passes over the drum. The positively charged paper attracts the negatively charged toner particles from the drum, effectively transferring the image onto the paper. To ensure complete transfer, the paper is often pre-treated with a slight negative charge on its surface, making it more attractive to the positively charged toner.

After the toner is transferred to the paper, it must be permanently fused to create a durable print. This is achieved through the fusing process, which involves passing the paper through a pair of heated rollers called the fuser assembly. The heat and pressure from the fuser assembly melt the plastic particles in the toner, causing them to bond with the fibers of the paper. The fuser rollers are typically coated with a non-stick material, such as Teflon, to prevent the toner from adhering to the rollers themselves.

The process doesn't end there; maintaining print quality and preventing residual toner from affecting subsequent prints is crucial. The photoreceptor drum is cleaned by a cleaning blade or brush that removes any leftover toner particles. Additionally, an erasing lamp or discharge LED neutralizes the remaining charge on the drum, preparing it for the next print cycle. This cleaning and erasing step ensure that each new print starts with a fresh, clean drum surface.

Laser printers are known for their high print quality, speed, and precision. The ability to produce sharp text and detailed images with a high level of accuracy is due to the fine control over the laser beam and the precision of the mechanical components. The resolution of a laser printer is typically measured in dots per inch (DPI), with higher DPI values indicating greater detail and clarity.

Color laser printers operate on the same basic principles as monochrome laser printers but with additional complexity. They use four separate photoreceptor drums and toner cartridges, each corresponding to one of the primary colors of the subtractive color model: cyan, magenta, yellow, and black (CMYK). The printing process for each color is sequential, and the paper passes through each drum and toner unit in a carefully coordinated manner. The individual color layers are precisely aligned, or registered, to create a full-color image. The fusing process then bonds all the layers together onto the paper.

The materials used in laser printing are critical to the overall performance and print quality. Toner composition, for instance, has seen significant advancements. Modern toners are engineered to have uniform particle size and shape, which improves the smoothness and consistency of printed images. The pigments in the toner provide the necessary color, while the plastic component ensures that the toner melts and bonds correctly during the fusing process.

Environmental considerations have also influenced the design and operation of laser printers. Manufacturers strive to reduce energy consumption through various means, such as improving the efficiency of the fuser assembly, which is typically the most energy-intensive part of the printer. Some printers use instant-on fusing technology, which allows the fuser to heat up quickly, reducing warm-up time and energy usage. Additionally, efforts are made to minimize the environmental impact of toner cartridges by designing them for recycling and reuse.

The integration of network connectivity and digital interfaces has further enhanced the functionality and convenience of laser printers. Modern printers often come equipped with Wi-Fi, Ethernet, and USB connections, allowing them to be easily integrated into home or office networks. This connectivity enables features such as remote printing, cloud-based document storage, and mobile printing, where users can send print jobs directly from their smartphones or tablets.

Advancements in printer software and firmware also play a significant role in optimizing performance and user experience. Printer drivers and management software allow users to customize print settings, monitor printer status, and manage print queues. Firmware updates can improve functionality, fix bugs, and add new features, extending the lifespan and versatility of the printer.

The evolution of laser printers continues with the development of new technologies aimed at enhancing print quality, speed, and energy efficiency. Innovations such as LED printing technology, which uses an array of light-emitting diodes instead of a laser, offer potential benefits in terms of cost, reliability, and compactness. Other emerging technologies focus on improving the resolution and color accuracy of printed images, expanding the capabilities of laser printers for professional and creative applications.

Chapter 14: Keys and Locks: Security through Mechanical Advantage

Keys and locks have been integral to human civilization for thousands of years, serving as fundamental tools for securing property and ensuring privacy. The evolution of keys and locks showcases the ingenuity and complexity of mechanical design, as well as the principles of physics and engineering that underlie their operation.

The history of locks dates back to ancient civilizations. The earliest known locks were created by the Egyptians around 4000 years ago. These early devices, known as pin tumbler locks, were made from wood and used a simple mechanism to secure doors. The key was a wooden peg with pins that aligned with holes in a bolt, allowing it to slide open when the correct key was inserted. This basic principle of aligning pins to open a lock remains central to many modern locking mechanisms.

As societies evolved, so did the need for more sophisticated security measures. The Greeks and Romans made significant advancements in lock technology, incorporating metal components and more intricate designs. Roman locks, for example, often featured wards, which were obstructions within the lock that matched specific notches on the key. Only a key with the correct notches could navigate the wards and operate the lock, adding an extra layer of security.

The principle of mechanical advantage is fundamental to the operation of keys and locks. Mechanical advantage refers to the use of simple machines to amplify force, making it easier to perform tasks. In the context of locks, mechanical advantage is achieved through the precise design of levers, springs, and pins that work together to secure or release the locking mechanism. When a key is inserted into a lock, it applies force to these components, moving them into alignment and allowing the lock to open.

The modern pin tumbler lock, invented by Linus Yale Sr. in the mid-19th century and refined by his son Linus Yale Jr., is one of the most widely used types of locks today. This lock consists of a cylindrical plug housed within a cylinder, or "casing." The plug has a keyway—a slot designed to accept a specific key profile. Within the plug and casing are sets of pins, typically in pairs, comprising a driver pin and a key pin. The driver pin rests above the key pin and is pushed down by a spring.

In the locked state, the pins extend into both the plug and the casing, preventing the plug from rotating. When the correct key is inserted, the uneven notches or cuts on the key align the key pins and driver pins at the shear line—the boundary between the plug and the casing. This alignment allows the plug to rotate freely, disengaging the locking mechanism and opening the lock. The precision required in manufacturing these components ensures that only a key with the exact cuts can align the pins correctly, providing security against unauthorized access.

Wafer tumbler locks, another common type, operate on a similar principle but use flat wafers instead of cylindrical pins. These locks are often found in automobile doors and filing cabinets. The key for a wafer tumbler lock has cuts or notches that correspond to the wafers within the lock. When the key is inserted, it pushes the wafers into alignment, allowing the lock to turn. While wafer tumbler locks are generally simpler and less secure than pin tumbler locks, they offer sufficient security for many applications.

Lever tumbler locks are yet another variation, commonly used in safes and older door locks. These locks contain a series of levers that must be lifted to a specific height by the key. Each lever has a gate, and only when all the gates are aligned with the bolt can the lock be opened. Lever tumbler locks can offer high security, especially when multiple levers are used, as the number of possible key combinations increases with each additional lever.

The development of combination locks introduced a different approach to security, relying on a sequence of numbers rather than a physical key. Combination locks use rotating dials connected to internal disks or cams. Each disk has a notch, and when the correct sequence of numbers is dialed, the notches align, allowing the lock to open. These locks are commonly used in safes and lockers, offering the advantage of keyless operation and eliminating the risk of lost or stolen keys.

Modern advancements in lock technology have led to the creation of electronic and smart locks, which offer enhanced security and convenience. Electronic locks use digital components and can be operated using keypads, biometric data, or wireless signals. For example, a keypad lock requires a user to enter a numerical code, which is verified by the lock's electronic circuitry before granting access. Biometric locks use fingerprints, retinal scans, or facial recognition to identify authorized users, providing a high level of security based on unique biological characteristics.

Smart locks take electronic locking technology a step further by integrating with home automation systems and the Internet of Things (IoT). These locks can be controlled remotely via smartphone apps, allowing users to lock or unlock doors from anywhere with an internet connection. Smart locks often feature additional security measures, such as real-time alerts and activity logs, enabling users to monitor access to their property. They can also be programmed to work with virtual assistants like Amazon Alexa or Google Assistant, providing voice-activated control and integration with other smart home devices.

Despite the advancements in lock technology, mechanical locks remain a cornerstone of physical security due to their reliability and simplicity.

The ongoing challenge is to balance security with ease of use. High-security locks, such as those used in safes and secure facilities, often incorporate multiple locking mechanisms to thwart unauthorized access attempts. For example, a high-security lock might combine a pin tumbler mechanism with a magnetic or electronic component, requiring both a physical key and an electronic code to open.

The materials used in lock construction also play a crucial role in their effectiveness and durability. High-quality locks are typically made from robust materials like brass, steel, and hardened alloys, which resist tampering and wear. The key itself is often made from brass or nickel-silver, chosen for their strength and resistance to corrosion. Advances in materials science have led to the development of locks with improved resistance to physical attacks, such as drilling, picking, and bumping.

Lock picking, a technique used to bypass locks without a key, exploits the mechanical principles of locks. Skilled lock pickers manipulate the internal components of a lock, such as pins or wafers, to align them at the shear line, simulating the action of a key. To counteract this, high-security locks may incorporate additional security features, such as mushroom pins, spool pins, or security pins, which make picking more difficult by creating false shear lines and increasing the skill required to pick the lock successfully.

Another common method of bypassing locks is lock bumping, which involves using a specially crafted "bump key" to apply force to the pins within the lock. The bump key is inserted into the lock and struck with a hammer or other tool, causing the pins to momentarily jump and align at the shear line, allowing the lock to turn. Manufacturers have developed various anti-bumping technologies to mitigate this risk, such as incorporating stronger springs, using uniquely shaped pins, and adding additional locking mechanisms.

In addition to physical security, the concept of key control is essential for maintaining security through locks. Key control involves managing the distribution and duplication of keys to ensure that only authorized individuals have access to them. High-security locks often come with patented key designs that can only be duplicated by the manufacturer or authorized dealers, preventing unauthorized key duplication and enhancing overall security.

The field of locks and security continues to evolve, driven by advancements in technology and the ongoing need for more effective security solutions. Research into new materials, innovative locking mechanisms, and enhanced electronic systems promises to further improve the reliability and security of locks in the future. As our understanding of security threats grows, so too does our ability to develop locks that provide robust protection against increasingly sophisticated attacks.

Chapter 15: The Balance of Bicycles: Gyroscopic Effects

Bicycles, a common sight on city streets and country roads alike, are marvels of mechanical engineering and physics. One of the most intriguing aspects of bicycles is their balance, particularly when in motion. This balance is largely attributed to gyroscopic effects, among other factors. Understanding how bicycles stay upright and maneuver through various terrains involves delving into the principles of gyroscopic motion, angular momentum, and the intricate interplay between these forces and the rider's inputs.

At the core of a bicycle's stability is the gyroscopic effect, which is a phenomenon that occurs when a rotating wheel generates angular momentum. Angular momentum is the quantity of rotation of an object and is dependent on the object's rotational speed and its moment of inertia, which is the resistance of an object to changes in its rotation. For a bicycle, the wheels act as gyroscopes, and as they spin, they generate angular momentum. This angular momentum contributes to the bicycle's ability to maintain balance while in motion.

To understand the gyroscopic effect, it is essential to consider the concept of precession. Precession is the tendency of a spinning object to react at a right angle to the direction of an applied force. When a bicycle wheel spins, it creates a gyroscopic effect that stabilizes the wheel. If an external force, such as a bump or a push, tries to tilt the bicycle to one side, the spinning wheels will react by generating a force at a right angle to the direction of the tilt. This reaction force helps counteract the tilt, contributing to the stability of the bicycle.

The stability provided by the gyroscopic effect becomes more pronounced at higher speeds. When a bicycle is moving quickly, the wheels spin faster, generating more angular momentum and thus a stronger gyroscopic effect. This is why it is easier to balance a bicycle at higher speeds; the increased angular momentum of the wheels helps resist tilting and enhances stability. Conversely, at lower speeds, the gyroscopic effect is weaker, making it more challenging to maintain balance.

However, the gyroscopic effect alone does not fully explain the balance of bicycles. The rider's inputs play a crucial role in maintaining stability and control. When a rider steers the bicycle, they are not only changing the direction of travel but also influencing the balance. Small adjustments in the handlebars can induce controlled lean angles, which help counteract any unwanted tilting. This process, known as countersteering, involves the rider momentarily turning the handlebars in the opposite direction of the desired turn to initiate a lean, then steering into the turn to maintain the lean and navigate the turn effectively.

Countersteering is an intuitive action that riders perform without consciously thinking about it. When a rider wants to turn left, they initially push the left handlebar forward, causing the front wheel to turn slightly to the right. This induces a lean to the left due to the shift in the bicycle's center of mass. Once the lean is established, the rider then steers into the turn, aligning the front wheel with the direction of the lean and guiding the bicycle through the turn. This dynamic interplay between steering and leaning is essential for maneuvering and maintaining balance on a bicycle.

Another critical factor in bicycle stability is the geometry of the bicycle itself, particularly the rake and trail of the front fork. The rake is the angle of the front fork relative to the vertical axis, while the trail is the horizontal distance between the point where the front wheel touches the ground and the point where the steering axis intersects the ground. The trail provides a self-correcting torque that helps align the front wheel with the direction of travel. When the bicycle begins to lean to one side, the trail causes the front wheel to steer into the lean, helping to restore balance.

The combination of gyroscopic effects, rider inputs, and bicycle geometry creates a complex system that allows for dynamic balance and control. The human body's proprioceptive feedback mechanisms—sensors in muscles and joints that provide information about body position and movement—also contribute to the rider's ability to maintain balance. Riders continually make small adjustments to their body position and steering to stay upright, often without conscious thought.

Additionally, the distribution of weight between the front and rear wheels plays a role in stability. Proper weight distribution helps ensure that the wheels maintain adequate contact with the ground, providing the necessary traction for steering and balance. Shifting weight forward or backward can affect the handling characteristics of the bicycle, influencing how it responds to steering inputs and external forces.

The interaction between the rider and the bicycle forms a feedback loop. As the rider makes adjustments to maintain balance, the bicycle's response provides sensory feedback, allowing the rider to make further corrections. This continuous process of adjustment and response is what enables riders to navigate various terrains and conditions, from smooth roads to rough trails, while maintaining stability.

Modern bicycles incorporate advanced materials and engineering techniques to enhance performance and stability. Lightweight yet strong materials like carbon fiber and aluminum alloys reduce the overall weight of the bicycle, making it easier to handle and control. Precision engineering ensures that components such as wheels, bearings, and frames are optimized for stability and efficiency.

Furthermore, advancements in technology have led to the development of smart bicycles equipped with sensors and electronic systems that assist with balance and control. Gyroscopic sensors, accelerometers, and electronic control units can monitor the bicycle's orientation and movement, providing real-time adjustments to enhance stability. Some bicycles are equipped with electronic steering dampers that reduce wobbling and improve handling, especially at high speeds.

Chapter 16: Scissors: Cutting through Shear Forces

Scissors, seemingly simple tools, are marvels of engineering designed to efficiently cut through various materials with precision. The cutting action of scissors relies on the application of shear forces, which are perpendicular to the surface being cut. Understanding the mechanics of scissors involves exploring the design principles, materials, and physical forces that enable them to perform their function effectively.

The basic structure of scissors consists of two blades joined together at a pivot point. Each blade has a sharp edge along one side, with a beveled cutting edge that tapers to a fine point. When the blades are brought together, the cutting edges intersect at an angle, forming a shearing action that slices through the material being cut. The pivot point allows the blades to pivot or rotate relative to each other, facilitating smooth and controlled cutting motions.

The cutting action of scissors is governed by the principles of shear force and deformation. Shear force occurs when two surfaces slide past each other in opposite directions, creating a parallel force that induces deformation or separation between the surfaces. In the case of scissors, the sharp edges of the blades apply shear force to the material being cut, causing it to deform and separate along the cut line. The precise alignment of the cutting edges and the sharpness of the blades are critical for achieving clean and accurate cuts.

The effectiveness of scissors depends on several factors, including blade geometry, blade material, and the mechanics of the cutting action. The angle of the cutting edges, known as the blade angle or bevel angle, determines how effectively the scissors can penetrate and shear through different materials. A sharper blade angle results in a finer cutting edge, which is suitable for cutting thin or delicate materials, while a broader angle provides more durability and strength for cutting thicker or tougher materials.

Blade material also plays a significant role in the performance of scissors. Common materials used for scissor blades include stainless steel, carbon steel, and various alloys. Stainless steel is favored for its corrosion resistance and durability, making it suitable for general-purpose scissors used in household and office settings. Carbon steel, on the other hand, offers superior hardness and edge retention, making it ideal for specialized scissors used in industrial or professional applications. Alloys such as titanium and cobalt are often used to enhance strength and wear resistance, particularly in high-performance scissors designed for cutting tough materials like Kevlar or carbon fiber.

The mechanics of the cutting action in scissors involve a combination of compression, tension, and shearing forces. As the blades come together, the material being cut is compressed between the cutting edges, creating a localized area of high pressure. This compression weakens the material, making it more susceptible to shearing forces. The sharpness and alignment of the cutting edges determine the efficiency of the shearing action, with sharper blades requiring less force to achieve a clean cut.

The design of scissors varies depending on the intended application and user preferences. Common types of scissors include household scissors, sewing scissors, kitchen shears, and specialty scissors for tasks like pruning, crafting, and medical procedures. Each type of scissors is optimized for specific cutting tasks, with variations in blade shape, size, and handle design to accommodate different materials and cutting techniques.

Household scissors, also known as general-purpose scissors, typically have symmetrical handles and straight blades with a moderate blade angle. They are suitable for a wide range of cutting tasks, including paper, cardboard, fabric, and light plastic. Sewing scissors, or tailor's shears, feature longer blades with a finer blade angle, allowing for precise cutting of fabric and thread. Kitchen shears, or kitchen scissors, often have serrated blades and additional features such as bottle openers and nutcrackers, making them versatile tools for food preparation.

Specialty scissors, such as pruning shears and craft scissors, are designed for specific applications that require specialized features. Pruning shears have curved blades and long handles for cutting branches and stems in gardening and landscaping. Craft scissors come in various shapes and sizes, including pinking shears for decorative cutting and scissors with serrated blades for cutting paper and cardboard with intricate patterns.

In addition to manual scissors, there are also powered scissors and shearing tools that use electric or pneumatic power to perform cutting tasks. Electric scissors, for example, are battery-powered or corded devices with motorized blades that provide increased cutting speed and efficiency, particularly for repetitive or heavy-duty cutting tasks. Pneumatic shears use compressed air to drive the cutting blades, offering a lightweight and portable solution for cutting materials like fabric, leather, and rubber.

The maintenance and care of scissors are essential for ensuring optimal performance and longevity. Regular sharpening of the blades helps maintain their sharpness and cutting efficiency. Sharpening can be done using sharpening stones, honing rods, or specialized sharpening tools designed for scissors. Proper storage, such as keeping scissors in a protective case or hanging them on a magnetic strip, helps prevent damage to the blades and maintains their alignment. Lubricating the pivot point with oil or grease reduces friction and ensures smooth operation of the scissors.

In summary, scissors are versatile cutting tools that rely on the application of shear forces to slice through various materials with precision. The design, materials, and mechanics of scissors are optimized for specific cutting tasks, with different types of scissors available to meet the needs of different users and applications. Whether for household chores, sewing projects, kitchen tasks, or specialized applications, scissors play an essential role in everyday life and demonstrate the principles of engineering and physics in action.

Chapter 17: Clocks and Timekeeping: Pendulums and Quartz Crystals

Clocks and timekeeping have been an integral part of human civilization for centuries, evolving from simple sundials to highly accurate atomic clocks. The development of reliable timekeeping devices has profoundly influenced various aspects of life, including navigation, science, industry, and daily activities. Two significant advancements in the history of clocks are the pendulum clock and the quartz crystal clock. These innovations leverage the principles of physics to measure time with remarkable precision.

Pendulum clocks, invented by Christiaan Huygens in 1656, were the first accurate timekeeping devices. A pendulum is a weight suspended from a pivot so that it can swing freely back and forth. The time it takes for a pendulum to complete one full swing, known as its period, is determined by the length of the pendulum and the acceleration due to gravity. Huygens discovered that a pendulum's period is independent of its amplitude, a property known as isochronism, which makes pendulums excellent timekeeping elements.

The pendulum clock operates on the principle of harmonic motion. When the pendulum swings, it moves through a regular, repeating cycle. The mechanism of the clock converts the pendulum's motion into the movement of the clock's hands. This is achieved through an escapement, a device that transfers energy to the pendulum to keep it swinging while also controlling the advance of the gear train that moves the clock hands.

The most common escapement in pendulum clocks is the anchor escapement, consisting of an anchor-shaped pallet and an escape wheel. As the pendulum swings, the pallets engage and release the teeth of the escape wheel, allowing it to advance by a fixed amount with each swing.

This controlled release of the escape wheel regulates the movement of the gears that drive the hour, minute, and second hands of the clock, ensuring accurate timekeeping.

The accuracy of pendulum clocks can be influenced by several factors, including temperature changes, which can affect the length of the pendulum, and air resistance, which can dampen the pendulum's motion. To mitigate these effects, clockmakers have developed various innovations. For instance, compensating pendulums, such as the gridiron pendulum, use rods made of different metals that expand and contract at different rates, minimizing the overall length change due to temperature variations. Additionally, clock cases can be designed to reduce the impact of air resistance and other environmental factors.

Despite their accuracy, pendulum clocks have limitations, particularly in terms of portability. The need for a stable, fixed location and sensitivity to external forces such as vibrations and tilting make them unsuitable for portable timekeeping. This limitation paved the way for the development of more versatile timekeeping technologies.

The invention of quartz crystal clocks in the early 20th century revolutionized timekeeping by providing a highly accurate and portable means of measuring time. Quartz clocks operate on the principle of piezoelectricity, a property of certain materials, such as quartz, that allows them to generate an electric charge in response to mechanical stress. Conversely, when an electric field is applied to a piezoelectric material, it induces mechanical vibrations at a precise frequency.

In a quartz clock, a small quartz crystal is cut and shaped to vibrate at a specific frequency when subjected to an electric field. This frequency, typically 32,768 Hz for most quartz clocks, is highly stable and can be used to regulate the clock's timekeeping mechanism. The crystal is housed in an oscillator circuit that generates a continuous oscillating electric signal at the crystal's resonant frequency.

The oscillating signal from the quartz crystal is fed into a series of frequency dividers, electronic circuits that reduce the frequency to a more manageable value, such as one pulse per second. These pulses drive a stepping motor or a digital display, advancing the clock's hands or digits to indicate the passage of time accurately.

The primary advantage of quartz clocks over pendulum clocks is their superior accuracy and stability. Quartz crystals are less affected by environmental factors such as temperature and humidity, which can cause mechanical clocks to drift. Additionally, quartz clocks are not sensitive to orientation or movement, making them ideal for portable and wearable timekeeping devices, such as wristwatches and mobile phones.

Quartz crystal clocks have undergone significant advancements since their invention. Improvements in manufacturing techniques have enabled the production of highly precise and stable quartz crystals, further enhancing the accuracy of these timekeeping devices. Modern quartz clocks can achieve accuracy within a few seconds per year, far surpassing the performance of most mechanical clocks.

The impact of accurate timekeeping extends beyond personal convenience. Precise time measurement is crucial for various scientific, industrial, and technological applications. For example, in telecommunications, accurate clocks are essential for synchronizing data transmission and ensuring the integrity of communication networks. In navigation, precise timekeeping allows for accurate determination of position using systems like GPS, which relies on synchronized signals from multiple satellites.

Atomic clocks represent the pinnacle of timekeeping accuracy. These clocks use the vibrations of atoms, typically cesium or rubidium, as their timekeeping element. The frequency of these atomic vibrations is incredibly stable, providing a timekeeping accuracy on the order of one second in millions of years. Atomic clocks are used to define the international standard for the second and provide the time signals that synchronize global positioning systems, telecommunications networks, and other critical infrastructure.

The development of cesium atomic clocks in the mid-20th century marked a significant milestone in timekeeping. These clocks use the transition frequency of cesium-133 atoms to define the second with unprecedented precision. The atoms are exposed to microwave radiation at a specific frequency, causing them to transition between energy states. By tuning the microwave frequency to match the cesium transition frequency, the atomic clock can maintain extremely accurate time.

Rubidium atomic clocks, while not as accurate as cesium clocks, offer high precision and are more compact and cost-effective, making them suitable for various practical applications. The principle of operation is similar to cesium clocks, but rubidium atoms are used instead. These clocks are often used in telecommunications and other applications where high accuracy is required, but the extreme precision of cesium clocks is not necessary.

The pursuit of even greater accuracy has led to the development of optical atomic clocks, which use transitions in optical frequencies rather than microwave frequencies. These clocks have the potential to achieve accuracies of one second in billions of years, further pushing the boundaries of timekeeping precision. Optical atomic clocks use laser-cooled ions or neutral atoms and advanced laser systems to achieve these remarkable levels of accuracy.

Chapter 18: The Elevator: Pulleys and Counterweights

Elevators are marvels of modern engineering, seamlessly blending principles of physics, mechanical ingenuity, and advanced technology to provide safe and efficient vertical transportation in buildings. At the core of their operation are the concepts of pulleys and counterweights, which have evolved over time to enhance the performance, safety, and energy efficiency of elevators. Understanding the intricate workings of elevators involves exploring these fundamental principles and their application in the complex systems that power modern elevators.

The basic principle behind an elevator is straightforward: a platform, or car, is moved vertically within a shaft to transport passengers or goods between different floors of a building. This vertical movement is achieved through a combination of mechanical systems, including electric motors, cables, pulleys, and counterweights. The pulleys and counterweights are essential components that enable the elevator to operate smoothly and efficiently.

Pulleys are simple machines consisting of a wheel with a grooved rim through which a rope or cable runs. They are used to change the direction of a force and can also provide a mechanical advantage, allowing a smaller force to lift a larger load. In the context of elevators, pulleys are employed in conjunction with electric motors to move the elevator car up and down the shaft.

The counterweight is a crucial element in the elevator system, designed to balance the weight of the elevator car and its load. By counterbalancing the elevator car, the counterweight reduces the amount of energy required to move the car, thereby enhancing the efficiency of the elevator system. The counterweight is typically a heavy mass located in the elevator shaft, connected to the car by a system of cables and pulleys.

The interaction between the pulleys and the counterweight is fundamental to the operation of the elevator. When the elevator car is at rest, the counterweight is positioned in such a way that it balances the car's weight. As the electric motor is activated to move the car, the pulleys redirect the force exerted by the motor, causing the car to ascend or descend within the shaft. The counterweight moves in the opposite direction of the car, maintaining the balance and reducing the load on the motor.

The use of counterweights provides several benefits to the elevator system. Firstly, it significantly reduces the energy consumption of the elevator. Since the counterweight offsets the weight of the car and its occupants, the motor only needs to provide enough power to overcome friction and inertia, rather than lifting the entire weight of the car. This energy-saving aspect is particularly important in tall buildings, where elevators must travel long distances and transport heavy loads frequently.

Secondly, counterweights contribute to the smooth operation and safety of elevators. By balancing the car's weight, the counterweight ensures that the car moves steadily and evenly, minimizing the risk of sudden jerks or stops that could cause discomfort or injury to passengers. The counterweight also helps to prevent the car from free-falling in the event of a mechanical failure or power outage. In such cases, the counterweight's mass can act as a brake, slowing down the car's descent and providing a crucial safety measure.

The design and placement of pulleys and counterweights in elevators have undergone significant advancements over time. Early elevators, such as those developed during the Industrial Revolution, used simple pulley systems with manual or steam-powered mechanisms. These early designs laid the groundwork for more sophisticated and reliable systems that emerged in the 20th century with the advent of electric motors and advanced engineering techniques.

Modern elevators typically use a combination of multiple pulleys and counterweights to achieve optimal performance. One common configuration is the traction elevator, which employs a motor-driven sheave (a specialized pulley with a grooved rim) to move the car and counterweight. The sheave is connected to the motor and rotates to wind or unwind the steel cables that support the car and counterweight. As the sheave turns, the car moves up or down the shaft, while the counterweight moves in the opposite direction.

The traction elevator system offers several advantages over older designs. The use of a sheave allows for greater control over the car's movement, enabling precise stops at each floor and smooth acceleration and deceleration. The steel cables, which are typically arranged in a loop over the sheave and connected to both the car and counterweight, provide a high degree of strength and durability, ensuring the safe operation of the elevator over long periods.

In addition to traction elevators, there are also hydraulic elevators, which use a different mechanism for vertical movement. Hydraulic elevators are powered by a hydraulic ram, a piston that moves within a cylinder filled with hydraulic fluid. The fluid is pumped into the cylinder to push the piston upward, lifting the car. To lower the car, the fluid is released back into the reservoir, allowing the piston to descend. While hydraulic elevators do not use counterweights, they still rely on pulleys to guide the car's movement and ensure stability.

The choice between traction and hydraulic elevators depends on various factors, including the building's height, the intended use of the elevator, and cost considerations. Traction elevators are preferred for taller buildings due to their energy efficiency and ability to travel longer distances quickly. Hydraulic elevators, on the other hand, are commonly used in low- to mid-rise buildings, where their simpler design and lower installation costs make them a practical choice.

Advancements in elevator technology continue to push the boundaries of what is possible, with innovations aimed at improving efficiency, safety, and passenger experience. One such advancement is the development of machine-room-less (MRL) elevators, which eliminate the need for a separate machine room by integrating the motor and control systems within the elevator shaft. MRL elevators use compact, gearless traction motors and advanced pulley systems to achieve the same performance as traditional elevators while saving space and reducing construction costs.

Another significant innovation is the use of regenerative drives, which capture the energy generated during the descent of the elevator car and convert it into electricity that can be fed back into the building's power grid. This technology not only enhances the energy efficiency of the elevator but also contributes to the overall sustainability of the building.

Safety remains a paramount concern in elevator design, and modern elevators are equipped with numerous safety features to protect passengers in the event of an emergency. These features include emergency brakes, which can engage if the car exceeds a certain speed, and redundant braking systems that ensure the car can be safely stopped even if one brake fails. Additionally, advanced monitoring and control systems continuously track the elevator's performance, detecting and diagnosing potential issues before they lead to malfunctions.

The passenger experience has also been a focus of elevator innovation. High-speed elevators, capable of traveling at speeds of up to 20 meters per second (45 miles per hour), are now common in skyscrapers, reducing travel time between floors and enhancing convenience. Elevators are also being designed with improved accessibility features, such as voice-activated controls, tactile buttons, and enhanced lighting, to accommodate passengers with disabilities.

The integration of smart technology into elevators is another exciting development. Smart elevators use sensors, data analytics, and connectivity to optimize performance and improve user experience. These elevators can adjust their operation based on real-time data, such as the number of passengers and their destinations, to reduce wait times and energy consumption. Additionally, smart elevators can be integrated with building management systems, enabling seamless coordination with other building functions, such as lighting, heating, and security.

Chapter 19: The Physics of Phones: Sound Waves and Signal Transmission

The development and widespread adoption of mobile phones represent one of the most significant technological advancements of the 20th and 21st centuries. These devices have revolutionized communication, enabling people to connect instantly over vast distances. The operation of mobile phones is a complex interplay of various physical principles, including sound waves, electromagnetic waves, signal processing, and transmission technologies.

At the core of phone communication is the conversion of sound waves into electrical signals and vice versa. Sound waves are mechanical vibrations that travel through a medium, such as air, water, or solid materials. These waves are characterized by their frequency, amplitude, and wavelength. When you speak into a phone, your voice generates sound waves that travel through the air and reach the phone's microphone.

The microphone in a phone is a transducer that converts sound waves into electrical signals. This conversion process involves a diaphragm inside the microphone that vibrates in response to the incoming sound waves. The diaphragm's vibrations cause changes in an electric circuit, producing an electrical signal that corresponds to the sound wave's frequency and amplitude. Modern phones typically use electret condenser microphones, which are sensitive, durable, and capable of capturing a wide range of frequencies.

Once the sound is converted into an electrical signal, the phone's electronic components process this signal. The signal is digitized, meaning it is converted from an analog signal (a continuous wave) into a digital signal (a series of discrete values). This digitization is achieved through a process called analog-to-digital conversion (ADC). During

ADC, the continuous sound wave is sampled at regular intervals, and each sample is assigned a numerical value based on the wave's amplitude at that point. The quality of the digitized signal depends on the sampling rate and the bit depth; higher values result in better sound quality.

The digital signal is then encoded and compressed to reduce the amount of data that needs to be transmitted. Encoding involves converting the digital signal into a format that can be efficiently transmitted over the network, while compression reduces the size of the data by removing redundant or less important information. Various encoding and compression algorithms are used in phones, such as Adaptive Multi-Rate (AMR) codec for voice signals.

After encoding and compression, the digital signal is modulated onto a carrier wave for transmission. Modulation is the process of varying one or more properties of a carrier wave (a high-frequency electromagnetic wave) in accordance with the digital signal. Common modulation techniques used in mobile communication include Frequency Modulation (FM), Amplitude Modulation (AM), Phase Modulation (PM), and Quadrature Amplitude Modulation (QAM). These techniques allow the encoded digital signal to be transmitted efficiently over long distances.

The modulated signal is then transmitted from the phone to the nearest cell tower via radio waves. Radio waves are a type of electromagnetic wave that can travel through the air and are suitable for wireless communication. The cell tower receives the radio signal and forwards it to the mobile network's core infrastructure, where it is routed to the recipient's phone. The core infrastructure includes various components such as base stations, switches, and gateways that manage the routing and switching of signals.

When the recipient's phone receives the signal, the process is reversed. The phone demodulates the incoming carrier wave, extracting the encoded digital signal. The signal is then decompressed and decoded back into its original digital form. Finally, a digital-to-analog converter (DAC) converts the digital signal back into an analog signal, which drives the phone's speaker to reproduce the original sound wave. This process allows the recipient to hear the sender's voice clearly.

In addition to voice communication, modern phones support data transmission for internet access, messaging, and multimedia applications. Data transmission involves similar principles but typically requires higher data rates and more advanced modulation and encoding techniques. Mobile data networks have evolved through several generations, each offering increased speeds and capabilities.

The first generation (1G) of mobile networks used analog signals for voice communication, with limited capacity and poor quality. The second generation (2G) introduced digital signals, enabling better voice quality, text messaging, and basic data services. Technologies such as Global System for Mobile Communications (GSM) and Code Division Multiple Access (CDMA) were standard in 2G networks.

The third generation (3G) marked a significant improvement in data transmission speeds, supporting more advanced applications such as video calls and mobile internet. 3G networks used technologies like Universal Mobile Telecommunications System (UMTS) and High-Speed Packet Access (HSPA) to achieve higher data rates and better performance.

The fourth generation (4G) further enhanced data speeds and network capacity, enabling seamless streaming, high-definition video calls, and other data-intensive applications. Long-Term Evolution (LTE) is the standard technology for 4G networks, providing significant improvements in speed, latency, and efficiency.

The fifth generation (5G) represents the latest advancement in mobile network technology, offering unprecedented data speeds, ultra-low latency, and massive connectivity. 5G networks use advanced technologies such as millimeter-wave frequencies, massive Multiple Input Multiple Output (MIMO) antennas, and beamforming to deliver high performance and support a wide range of applications, from augmented reality to autonomous vehicles.

Signal transmission in mobile networks involves a combination of wired and wireless technologies. The wireless component includes the transmission of signals between phones and cell towers using radio waves, while the wired component involves the routing of signals through fiber-optic cables, switches, and routers within the core network. Fiber-optic cables are essential for high-speed data transmission, as they use light waves to carry signals over long distances with minimal loss and interference.

Interference and signal degradation are critical challenges in mobile communication. Various factors can affect signal quality, including physical obstructions (buildings, trees), electromagnetic interference from other devices, and environmental conditions (weather, terrain). Mobile networks use several techniques to mitigate these issues, such as:

1. **Frequency Reuse**: Allocating different frequencies to adjacent cells to minimize interference.
2. **Cell Splitting**: Dividing large cells into smaller ones to increase capacity and reduce interference.
3. **Adaptive Modulation and Coding (AMC)**: Adjusting modulation and coding schemes based on signal quality to optimize data transmission.
4. **Error Correction**: Using techniques like forward error correction (FEC) to detect and correct errors in transmitted

data.

5. **Diversity Techniques**: Implementing methods like spatial diversity (using multiple antennas) and frequency diversity (transmitting over different frequencies) to improve signal reliability.

The evolution of mobile phones has also been driven by advancements in semiconductor technology, battery technology, and software development. Modern smartphones are powerful computing devices equipped with high-resolution displays, multiple cameras, sensors (accelerometers, gyroscopes, proximity sensors), and advanced operating systems. These features enable a wide range of applications beyond voice communication, including photography, gaming, navigation, and health monitoring.

Chapter 20: Vacuum Cleaners: Suction and Air Flow

Vacuum cleaners are ubiquitous household appliances, their fundamental purpose being the removal of dust, dirt, and debris from floors, carpets, and other surfaces. Understanding how vacuum cleaners work involves delving into the principles of suction and air flow, as well as the various mechanical and electrical components that make these devices effective. Over the years, vacuum cleaners have evolved significantly, incorporating advanced technologies to improve efficiency, suction power, and user convenience.

The core principle behind a vacuum cleaner is the creation of a partial vacuum inside the device. A vacuum is essentially a space devoid of matter, including air. By creating a partial vacuum, the vacuum cleaner generates a pressure difference between the inside of the device and the external environment. This pressure difference causes air to rush into the vacuum cleaner, carrying dust and debris along with it.

The key component responsible for creating suction in a vacuum cleaner is the motor. The motor drives a fan, which rapidly spins to create a low-pressure area inside the vacuum cleaner. As the fan blades rotate, they push air out of the vacuum cleaner, reducing the air pressure inside. This action generates a suction force that pulls air from the outside environment into the vacuum cleaner through its intake port. The speed and power of the motor directly influence the strength of the suction, making it a critical factor in the vacuum cleaner's performance.

Air flow is another crucial aspect of vacuum cleaner operation. It refers to the movement of air from the point of intake, through the vacuum cleaner, and out through the exhaust. The design and efficiency of the air flow path significantly affect the vacuum cleaner's ability to capture and retain dust and debris. A well-designed air flow path ensures that air moves smoothly through the vacuum cleaner, minimizing turbulence and resistance, which can reduce suction power and overall efficiency.

The journey of air through a vacuum cleaner can be broken down into several stages:

1. **Intake**: When the vacuum cleaner is turned on, the motor and fan create a low-pressure area inside the vacuum cleaner. This causes air to rush in through the intake port, typically located at the bottom of the vacuum cleaner. The intake port is connected to a nozzle or a brush head, which comes into contact with the surface being cleaned. The design of the nozzle or brush head is critical for effective cleaning, as it must ensure good contact with the surface and facilitate the efficient transfer of dirt and debris into the vacuum cleaner.

2. **Filtration**: As the air, along with dust and debris, enters the vacuum cleaner, it passes through one or more filters. The purpose of the filters is to trap dust, dirt, and other particles, preventing them from being recirculated back into the air. Vacuum cleaners typically use multiple stages of filtration to ensure thorough cleaning. The first stage is usually a pre-filter, which captures larger particles and protects the main filter from clogging. The main filter, often made of high-efficiency particulate air (HEPA) material, captures smaller particles, including allergens and fine dust. HEPA filters are particularly effective, trapping particles as small as 0.3 microns with an efficiency of 99.97%.

3. **Dirt Collection**: After passing through the filters, the cleaned air continues its journey through the vacuum cleaner. The dust and debris collected by the filters are deposited in a dust bag or a dustbin. Traditional vacuum cleaners use disposable dust

bags, which need to be replaced when full. Modern vacuum cleaners, however, often use bagless designs with removable dustbins, which can be emptied and reused. Bagless vacuum cleaners typically use cyclonic separation to remove dust and debris from the air. In cyclonic vacuum cleaners, the incoming air is forced into a cyclone or a vortex. The centrifugal force generated by the spinning air causes dust and debris to be flung outward, where they are collected in the dustbin.

4. **Exhaust**: After passing through the filtration system, the cleaned air is expelled from the vacuum cleaner through the exhaust port. The design of the exhaust system is important to minimize noise and ensure that any remaining fine particles are not released back into the environment. Some vacuum cleaners use additional filters at the exhaust stage to further purify the air before it is released.

Several factors influence the performance of a vacuum cleaner, including suction power, air flow, and the efficiency of the filtration system. Suction power is typically measured in units such as pascals (Pa) or inches of water lift. Air flow is measured in cubic feet per minute (CFM) or liters per second (L/s). The balance between suction power and air flow is crucial for effective cleaning. High suction power is necessary to lift dirt and debris from surfaces, while adequate air flow ensures that the particles are transported through the vacuum cleaner and captured by the filters.

Different types of vacuum cleaners are designed to address specific cleaning needs and environments. Here are some common types of vacuum cleaners and their unique features:

1. **Upright Vacuum Cleaners**: Upright vacuum cleaners are popular for cleaning large carpeted areas. They typically have a powerful motor and a wide cleaning path, making them

efficient for covering large surfaces quickly. Upright vacuum cleaners often come with adjustable brush heights to accommodate different carpet thicknesses and hard floors.

2. **Canister Vacuum Cleaners**: Canister vacuum cleaners consist of a separate canister unit connected to a hose and a cleaning wand. They offer greater flexibility and maneuverability compared to upright vacuum cleaners, making them ideal for cleaning stairs, furniture, and hard-to-reach areas. Canister vacuum cleaners are also well-suited for hard floors and low-pile carpets.

3. **Stick Vacuum Cleaners**: Stick vacuum cleaners are lightweight and easy to handle, making them convenient for quick clean-ups and small spaces. They are typically cordless, powered by rechargeable batteries, and are less powerful than upright or canister vacuum cleaners. Stick vacuum cleaners are ideal for light cleaning tasks and can be easily stored in small spaces.

4. **Robotic Vacuum Cleaners**: Robotic vacuum cleaners are autonomous devices that navigate and clean floors on their own. They use sensors and algorithms to map and avoid obstacles, making them suitable for automated daily cleaning. While not as powerful as traditional vacuum cleaners, robotic vacuum cleaners are convenient for maintaining cleanliness with minimal effort.

5. **Handheld Vacuum Cleaners**: Handheld vacuum cleaners are compact and portable, designed for quick spot cleaning and small messes. They are ideal for cleaning cars, upholstery, and tight spaces. Handheld vacuum cleaners are usually cordless and powered by rechargeable batteries.

6. **Wet/Dry Vacuum Cleaners**: Wet/dry vacuum cleaners are versatile machines capable of handling both liquid spills and dry debris. They are commonly used in workshops, garages,

and construction sites. Wet/dry vacuum cleaners have robust motors and durable construction to withstand demanding cleaning tasks.

Advancements in vacuum cleaner technology have led to the development of features that enhance performance, convenience, and user experience. Some of these features include:

1. **Cyclonic Technology**: Cyclonic vacuum cleaners use cyclonic separation to remove dust and debris from the air. The spinning air creates a cyclone, which separates particles based on their size and weight. This technology prevents clogging and maintains consistent suction power.
2. **HEPA Filtration**: High-efficiency particulate air (HEPA) filters are capable of capturing extremely small particles, including allergens, pollen, and pet dander. HEPA filters are essential for maintaining indoor air quality and are particularly beneficial for allergy sufferers.
3. **Cordless Operation**: Cordless vacuum cleaners are powered by rechargeable batteries, providing greater mobility and convenience. Advances in battery technology, such as lithium-ion batteries, have improved the runtime and power of cordless vacuum cleaners.
4. **Smart Features**: Modern vacuum cleaners are increasingly equipped with smart features, such as Wi-Fi connectivity, app control, and voice commands. These features allow users to schedule cleaning sessions, monitor performance, and receive maintenance alerts remotely.
5. **Adjustable Suction**: Some vacuum cleaners offer adjustable suction settings to accommodate different cleaning tasks and surfaces. Lower suction settings are suitable for delicate fabrics and curtains, while higher settings are ideal for deep cleaning carpets and rugs.

6. **Motorized Brush Heads**: Motorized brush heads enhance the cleaning performance of vacuum cleaners by agitating carpet fibers and dislodging embedded dirt. These brush heads are particularly effective for deep cleaning carpets and removing pet hair.

7. **Allergen Seals**: Some vacuum cleaners are designed with allergen seals and advanced filtration systems to prevent dust and allergens from escaping back into the air. These features are crucial for maintaining a clean and healthy indoor environment.

The future of vacuum cleaner technology continues to hold promise for further innovations that will enhance cleaning efficiency, user convenience, and environmental sustainability. Developments in artificial intelligence and robotics are likely to result in more advanced and autonomous cleaning solutions, while improvements in battery technology and energy efficiency will contribute to longer runtimes and reduced environmental impact.

Chapter 21: The Dynamics of Drawers: Sliding and Rolling Friction

The dynamics of drawers, whether they are part of a kitchen cabinet, a dresser, or an office desk, involve a fascinating interplay of mechanical principles, primarily focusing on sliding and rolling friction. Understanding how drawers work requires a deep dive into the physics of motion, material interactions, and the engineering solutions developed to ensure smooth, efficient, and durable operation. This comprehensive exploration will cover the principles of friction, types of drawer mechanisms, materials used in drawer construction, and innovations in drawer design.

At the heart of drawer operation is the concept of friction, which is the resistance to motion that occurs when two surfaces interact. There are two primary types of friction relevant to drawers: sliding friction and rolling friction.

Sliding friction, also known as kinetic friction, occurs when two surfaces slide against each other. In the context of a drawer, sliding friction is the resistance encountered when the drawer moves along its track or guide. The magnitude of sliding friction depends on the nature of the surfaces in contact, the force pressing them together (normal force), and the coefficient of friction, which is a measure of how easily one surface slides over another. The coefficient of friction varies depending on the materials and surface finishes involved.

Rolling friction, on the other hand, occurs when an object rolls over a surface. In the case of drawers, rolling friction is encountered in systems that use wheels or rollers to facilitate movement. Rolling friction is generally much lower than sliding friction, making it an attractive option for reducing the effort required to open and close drawers. The magnitude of rolling friction depends on factors such as the material and diameter of the wheels or rollers, the surface they roll on, and the load being carried.

Drawer mechanisms can be broadly categorized into two types: those that rely primarily on sliding friction and those that utilize rolling elements. Each type has its own set of characteristics, advantages, and applications.

Traditional drawer mechanisms often rely on sliding friction. These drawers typically use wooden or metal tracks that guide the drawer as it moves in and out. The interaction between the drawer and the tracks generates sliding friction, which must be overcome to operate the drawer. To reduce friction and facilitate smoother movement, these mechanisms may incorporate lubricants, such as wax, oil, or grease, applied to the contact surfaces. Additionally, modern sliding mechanisms often use materials with low coefficients of friction, such as plastics or polymers, to minimize resistance.

Metal or ball-bearing drawer slides are a significant advancement over traditional wooden slides. Ball-bearing slides utilize small metal balls contained within a track to convert sliding friction into rolling friction. As the drawer moves, the balls roll along the tracks, significantly reducing friction and making the drawer operation much smoother and more efficient. Ball-bearing slides are particularly effective in supporting heavier loads and are commonly used in applications where durability and ease of use are paramount, such as in office furniture, tool chests, and kitchen cabinets.

Another innovative type of drawer mechanism is the roller drawer slide. Roller slides use cylindrical rollers instead of ball bearings. These rollers are typically made of durable materials like nylon or metal and roll along tracks as the drawer moves. Roller slides combine the low friction benefits of rolling mechanisms with the simplicity of traditional sliding systems. They are often used in applications where a balance between cost, durability, and smooth operation is desired.

The materials used in drawer construction play a crucial role in the performance and longevity of the drawer mechanism. Traditional drawers were often made entirely of wood, with wooden tracks and runners. While wood offers a warm aesthetic and is relatively easy to work with, it also presents challenges, such as susceptibility to wear, changes in humidity causing swelling or shrinking, and higher friction compared to modern materials.

In contemporary drawer design, a variety of materials are used to enhance performance and durability. Metals, such as steel and aluminum, are commonly used for drawer slides and tracks due to their strength, rigidity, and resistance to wear. Metal slides can be precisely engineered to ensure consistent performance and can support heavier loads without deforming.

Plastics and polymers, such as nylon and Teflon, are also frequently used in drawer mechanisms. These materials have low coefficients of friction, excellent wear resistance, and can be easily molded into complex shapes. Plastic components are often used in conjunction with metal slides to reduce noise and provide a smoother sliding action.

Composite materials, combining the benefits of different substances, are also gaining popularity in drawer construction. For example, some drawer slides use a combination of metal tracks and plastic rollers or bearings to achieve a balance between strength, low friction, and quiet operation.

Innovations in drawer design have led to the development of advanced features that enhance functionality, user experience, and safety. One such innovation is the soft-close mechanism, which uses hydraulic or pneumatic dampers to slow down the closing action of the drawer.

This prevents the drawer from slamming shut, reducing noise, and preventing damage to the drawer and its contents. Soft-close mechanisms also enhance safety by reducing the risk of fingers getting caught in the closing drawer.

Another innovative feature is the push-to-open mechanism, which allows drawers to be opened with a simple push on the front panel. This mechanism typically uses spring-loaded catches or electronic actuators to release the drawer from its closed position. Push-to-open drawers provide a sleek, handle-free design and are particularly popular in modern, minimalist interiors.

Full-extension drawer slides are another advancement that allows drawers to be pulled out completely, providing full access to the contents. Traditional drawer slides often only allow partial extension, making it difficult to reach items at the back. Full-extension slides use telescoping mechanisms that ensure the drawer can be fully extended while maintaining stability and support.

Undermount drawer slides are a design innovation that hides the sliding mechanism underneath the drawer, providing a clean and uncluttered appearance. These slides often incorporate features such as soft-close and full-extension capabilities, combining functionality with aesthetic appeal. Undermount slides are commonly used in high-end cabinetry and furniture design.

Ergonomics is also a critical consideration in drawer design. The height, depth, and ease of access to drawers can significantly impact user comfort and efficiency. Ergonomically designed drawers are often placed at convenient heights to reduce bending and reaching, and may include dividers, organizers, and pull-out trays to enhance accessibility and organization.

The dynamics of drawers extend beyond residential furniture and are also crucial in industrial and commercial applications. In these settings, drawers are often subjected to heavier loads and more demanding usage conditions. Heavy-duty drawer slides, designed to support substantial weight and frequent operation, are essential in environments such as manufacturing plants, laboratories, and commercial kitchens. These slides are typically made from high-strength materials and incorporate reinforced construction to ensure durability and reliability.

In summary, the dynamics of drawers encompass a complex interplay of mechanical principles, material properties, and engineering innovations. Sliding and rolling friction are fundamental forces that dictate the performance and efficiency of drawer mechanisms. Advances in materials and design have led to the development of sophisticated drawer systems that offer smooth operation, enhanced functionality, and improved aesthetics. Whether in residential, commercial, or industrial settings, the continued evolution of drawer technology promises to deliver even greater convenience, durability, and user satisfaction.

Chapter 22: Bridges: Tension, Compression, and Load Distribution

Bridges are one of the most fascinating and vital components of human infrastructure, enabling the passage over obstacles such as rivers, valleys, or other roads. The design and construction of bridges involve a deep understanding of various physical principles, particularly the forces of tension, compression, and load distribution. This detailed exploration will cover the fundamental principles behind these forces, the different types of bridges, the materials used in bridge construction, and how modern engineering techniques ensure the durability and stability of these crucial structures.

To understand how bridges function, it is essential to grasp the concepts of tension and compression. These are the primary forces that act on any structure. Tension is the force that attempts to elongate or stretch a material, while compression is the force that attempts to shorten or squeeze a material. Effective bridge design requires balancing these forces to ensure that the structure can withstand the loads it will encounter without failing.

Load distribution is another critical concept in bridge engineering. Bridges must support various loads, including the dead load (the weight of the bridge itself), the live load (the weight of vehicles, pedestrians, and other transient factors), and environmental loads (such as wind, earthquakes, and temperature variations). The ability of a bridge to distribute these loads effectively across its structure is crucial for its stability and longevity.

There are several types of bridges, each utilizing different methods to manage tension, compression, and load distribution. The most common types include beam bridges, arch bridges, suspension bridges, cable-stayed bridges, and truss bridges.

Beam bridges are the simplest and oldest form of bridge design. They consist of a horizontal beam supported at both ends. The weight of the load on the bridge causes the beam to bend, creating tension on the bottom side of the beam and compression on the top side. The ability of a beam bridge to carry a load depends on the material's tensile and compressive strengths, as well as the beam's length and cross-sectional shape. Modern beam bridges often use materials like steel or reinforced concrete to maximize strength and durability.

Arch bridges utilize the principle of compression to distribute loads. An arch bridge transfers the weight of the load from the bridge deck to the abutments (the supports at either end of the bridge) through the curved structure of the arch. The arch shape naturally directs the forces downward and outward, efficiently distributing the load across the entire structure. The materials used for arch bridges must be capable of withstanding significant compressive forces, with stone, brick, and concrete being common choices historically, and reinforced concrete and steel being prevalent in modern designs.

Suspension bridges are notable for their ability to span very long distances. They utilize cables suspended from towers to support the bridge deck. The cables experience tension as they bear the weight of the load, while the towers experience compression. The main cables are anchored securely at both ends of the bridge, transferring the load to the ground. Suspension bridges are particularly effective at distributing loads across their length, making them ideal for spanning wide gaps such as large rivers or bays. The Golden Gate Bridge in San Francisco is a famous example of a suspension bridge.

Cable-stayed bridges are similar to suspension bridges but use a different method to support the bridge deck. Instead of main cables running the length of the bridge, cable-stayed bridges use multiple cables connected directly from the towers to the bridge deck. These cables fan out in a pattern and are under tension, while the towers experience compression. Cable-stayed bridges provide excellent load distribution and can be designed in various configurations, such as radial or harp patterns. The Millau Viaduct in France is a notable example of a cable-stayed bridge.

Truss bridges use a framework of interconnected triangles (trusses) to distribute loads. The triangular shapes are inherently strong and efficient at managing both tension and compression. The trusses distribute the forces throughout the structure, minimizing the stress on any single point. Truss bridges can be designed in various configurations, such as the Warren truss, Pratt truss, or Howe truss, each with its own method of distributing loads. These bridges are often constructed from steel due to its high tensile and compressive strength, making truss bridges suitable for both short and long spans.

The materials used in bridge construction are crucial for ensuring the structure can handle the forces of tension and compression effectively. Historically, materials such as stone, wood, and iron were used in bridge construction. Stone is excellent at withstanding compression, making it ideal for arch bridges, but it is heavy and less suitable for tension. Wood is relatively strong in both tension and compression but is susceptible to decay and less durable over long periods.

With the advent of the industrial revolution, iron and steel became prominent in bridge construction. Steel, in particular, has a high strength-to-weight ratio and excellent tensile and compressive properties, making it ideal for a wide range of bridge designs, including beam, truss, suspension, and cable-stayed bridges. Reinforced concrete, which combines the compressive strength of concrete with the tensile strength of steel reinforcement, is also widely used in modern bridge construction. This composite material can be molded into various shapes and provides excellent durability and load-bearing capacity.

Modern bridge engineering also incorporates advanced techniques to enhance stability, durability, and safety. One such technique is the use of prestressed concrete, where steel tendons are tensioned before the concrete is poured. Once the concrete hardens, the tendons are released, putting the concrete into compression. This method increases the load-bearing capacity of the concrete and reduces the risk of cracking.

The design of bridge foundations is another critical aspect of ensuring stability. Foundations must transfer the loads from the bridge to the ground, and their design depends on the geological conditions of the site. Common foundation types include shallow foundations, such as spread footings, and deep foundations, such as piles and caissons. Deep foundations are used when the surface soil is not strong enough to support the loads, requiring the foundation to extend down to more stable soil or rock layers.

The use of computer modeling and simulation has revolutionized bridge design and construction. Engineers can create detailed models of bridges, analyze the forces and stresses acting on the structure, and optimize the design for maximum efficiency and safety. These tools allow for precise calculations of load distribution, material behavior, and the impact of environmental factors such as wind and earthquakes.

Another important consideration in bridge design is the impact of dynamic loads, such as traffic and wind. Bridges must be designed to accommodate not only static loads (the weight of the bridge itself and any stationary objects) but also dynamic loads, which can vary in magnitude and direction. Traffic loads, for example, can create vibrations and oscillations that must be managed to prevent structural fatigue. Wind loads can induce sway and torsional forces, especially in long-span bridges. Engineers use techniques such as dampers, tuned mass dampers, and aerodynamic profiling to mitigate these dynamic effects.

Maintenance and inspection are crucial for ensuring the long-term safety and functionality of bridges. Regular inspections identify potential issues such as corrosion, cracking, and fatigue, allowing for timely repairs and maintenance. Advanced technologies, such as drones and sensors, are increasingly used for bridge inspection, providing detailed data and reducing the need for manual inspections in hazardous conditions.

Chapter 23: Remote Controls: Infrared Technology

Remote controls have become an indispensable part of modern life, providing convenient and wireless operation of a variety of electronic devices, including televisions, air conditioners, and sound systems. The underlying technology that makes remote controls work seamlessly is often infrared (IR) technology. This technology has been refined over the decades to become a reliable and cost-effective means of transmitting signals from a remote control to a device.

Infrared technology relies on the transmission of infrared light, which is a type of electromagnetic radiation with wavelengths longer than visible light but shorter than microwaves. Infrared light is invisible to the human eye but can be detected by electronic sensors. The primary principle behind IR remote controls is the emission and reception of infrared light signals, which are modulated to carry information.

A typical infrared remote-control system consists of two main components: the transmitter (remote control) and the receiver (electronic device). The transmitter generates an infrared light signal that is modulated to encode the command being sent. The receiver detects this signal and demodulates it to interpret the command.

The transmitter in a remote control is equipped with an infrared LED (light-emitting diode). When a button on the remote control is pressed, an electrical circuit is completed, and the remote control's microcontroller activates the infrared LED. The LED emits infrared light pulses that are modulated to correspond to a specific command, such as turning on the TV or changing the channel.

Modulation is a critical aspect of infrared communication. It involves varying a property of the infrared light wave, such as its amplitude, frequency, or phase, to encode the information. The most commonly used modulation technique in IR remote controls is pulse-width modulation (PWM), where the duration of each pulse corresponds to a specific bit of information. The pulses are typically transmitted at a carrier frequency of around 38 kHz, which helps distinguish the signal from ambient infrared light sources like sunlight or indoor lighting.

When the modulated infrared light reaches the electronic device, it is detected by an infrared receiver module. This module usually consists of a photodiode or phototransistor that converts the incoming infrared light back into an electrical signal. The receiver is tuned to the carrier frequency used by the transmitter, allowing it to filter out other infrared light sources and focus on the modulated signal.

The electrical signal from the photodiode is then passed to a demodulator, which extracts the modulated information by decoding the pulse durations. This information is then processed by the device's microcontroller, which interprets the command and executes the corresponding action, such as increasing the volume or changing the input source.

Encoding schemes are crucial for ensuring that the transmitted signal can be accurately interpreted by the receiver. Different manufacturers use various encoding protocols, such as the RC-5 protocol developed by Philips, the NEC protocol, and the Sony SIRC protocol. These protocols define the specific patterns of pulses and spaces that correspond to different commands. For example, in the RC-5 protocol, each command is represented by a 14-bit code, including a start bit, control bits, and the command itself. The use of unique codes for each button press ensures that the correct command is executed by the device.

One of the advantages of infrared technology is its simplicity and cost-effectiveness. Infrared LEDs and receivers are inexpensive and easy to integrate into electronic devices, making them a popular choice for remote controls. The line-of-sight nature of infrared communication, where the transmitter and receiver must be directly aligned, helps prevent interference from other devices and ensures that the commands are directed to the intended device.

However, infrared technology also has some limitations. The need for a direct line of sight means that obstacles between the remote control and the device can block the signal, reducing reliability in certain situations. Additionally, infrared signals can be affected by strong ambient light, such as sunlight, which can interfere with communication. The range of infrared remote controls is typically limited to around 5 to 10 meters, which can be restrictive in larger spaces.

Despite these limitations, infrared remote controls have been widely adopted due to their effectiveness and ease of use. Over the years, various improvements have been made to enhance their performance. For example, some modern remote controls use multiple infrared LEDs to increase the signal strength and widen the transmission angle, reducing the need for precise aiming. Others incorporate advanced encoding schemes that improve resistance to interference and allow for more reliable operation.

Looking to the future, infrared technology continues to evolve, with developments aimed at addressing its limitations and expanding its capabilities. One area of innovation is the integration of infrared communication with other wireless technologies, such as Bluetooth and RF (radio frequency). This hybrid approach allows for more flexible and reliable remote-control solutions that can operate without the strict line-of-sight requirement of traditional infrared systems.

Another promising development is the use of infrared communication in smart home systems. Infrared remote controls can be integrated with smart home hubs, enabling users to control multiple devices through a single interface, such as a smartphone app or voice command. This integration can provide a seamless and convenient user experience, where infrared signals are relayed by the smart hub to the appropriate devices.

Furthermore, advancements in infrared sensors and LEDs continue to improve the efficiency and range of infrared communication. New materials and technologies are being explored to create more sensitive photodetectors and more powerful infrared emitters, which could enhance the performance of remote controls and other infrared-based systems.

Chapter 24: The Stability of Shelves: Torque and Equilibrium

The stability of shelves is a fundamental aspect of both practical and theoretical interest, blending concepts from physics, engineering, and everyday life. At the heart of understanding how shelves remain stable—or why they might fail—is the study of torque and equilibrium.

To begin with, the concept of torque is essential for understanding the stability of shelves. Torque, also known as the moment of force, is a measure of the rotational force acting on an object. It is calculated as the product of the force applied and the distance from the point of application to the axis of rotation, known as the lever arm. Mathematically, torque (τ) is expressed as:

$$\tau = F \times r \times \sin(\vartheta)$$

where F is the magnitude of the force, r is the length of the lever arm, and ϑ is the angle between the force vector and the lever arm. For shelves, torque comes into play when forces, such as the weight of the items placed on the shelf, create a rotational effect around the points where the shelf is supported.

Equilibrium is the state in which all forces and torques acting on an object are balanced, resulting in no net motion. For a shelf to be stable, it must be in both translational and rotational equilibrium. Translational equilibrium means that the sum of all forces acting on the shelf is zero, ensuring that the shelf does not move linearly. Rotational equilibrium means that the sum of all torques acting on the shelf is zero, preventing the shelf from rotating around its supports.

Consider a simple wall-mounted shelf supported by brackets. When the shelf is loaded with items, each item exerts a downward force due to gravity, creating a torque around the points where the shelf is attached to the wall. The stability of the shelf depends on the distribution of these torques and the ability of the brackets to counteract them.

The design and material of the shelf and its supports play a critical role in its stability. Shelves are typically made from materials such as wood, metal, or plastic, each with its own characteristics. Wood is a common choice for shelves due to its aesthetic appeal and ease of fabrication. However, wood has a relatively low strength-to-weight ratio compared to metals, which means it can deform under heavy loads. Metal shelves, often made from steel or aluminum, offer higher strength and rigidity but can be more expensive and heavier. Plastic shelves are lightweight and resistant to moisture but may lack the strength required for heavy loads.

The choice of material affects the shelf's ability to resist bending and deformation under load. The material's Young's modulus, a measure of stiffness, is crucial in determining how much the shelf will bend under a given load. For instance, steel has a higher Young's modulus than wood, meaning a steel shelf will bend less than a wooden shelf of the same dimensions under the same load.

In addition to material properties, the geometric design of the shelf and its supports is vital for stability. The thickness and width of the shelf influence its ability to bear loads without excessive bending. Thicker shelves are generally more stable and less prone to bending but add more weight to the overall structure. The length of the shelf also plays a role; longer shelves are more likely to experience significant bending and require additional support to remain stable.

The type and placement of supports are equally important. Brackets, which are the most common type of support, come in various designs, from simple L-shaped brackets to more complex designs that distribute loads more evenly. The placement of brackets affects the load distribution and the torques acting on the shelf. Brackets should be positioned to minimize the unsupported span of the shelf and distribute the load as evenly as possible.

When analyzing the stability of a shelf, it is crucial to consider the forces and torques acting on it. The weight of the shelf itself and any items placed on it create a downward force, which must be counteracted by the upward forces exerted by the supports. If these forces are not balanced, the shelf may either tip over or detach from its supports. Additionally, the torque created by the weight of the items must be balanced by the counteracting torque provided by the supports.

Let's consider a practical example. Imagine a shelf supported by two brackets placed equidistant from the ends of the shelf. If an object is placed in the center of the shelf, it creates a downward force directly between the brackets, generating equal and opposite torques on each bracket. This setup ensures that the shelf remains in equilibrium and stable. However, if the object is placed closer to one bracket, the torque on that bracket increases, potentially causing the shelf to tip or the bracket to fail if it is not designed to handle the additional load.

External factors such as vibrations, impact forces, and environmental conditions can also affect shelf stability. For instance, in regions prone to earthquakes, shelves must be designed to withstand seismic forces that can cause sudden and intense vibrations. The use of flexible materials, secure anchoring systems, and additional bracing can help enhance the stability of shelves in such environments.

In modern engineering and design, various innovations have been developed to enhance the stability and functionality of shelves. Adjustable shelving systems, for example, allow users to change the position of supports to optimize load distribution and maintain stability. Floating shelves, which appear to be unsupported, use concealed brackets and internal supports to achieve a sleek appearance while maintaining stability.

Shelving units with integrated reinforcement, such as cross-bracing or tension cables, can provide additional stability by distributing loads more effectively and preventing deformation. In industrial and commercial settings, heavy-duty shelving systems often use steel frames and reinforced beams to support substantial loads safely.

The use of advanced materials and fabrication techniques has also contributed to improved shelf stability. Composite materials, which combine the strength of metals with the lightness of plastics, offer high strength-to-weight ratios and excellent resistance to bending and deformation. Techniques such as 3D printing and precision machining allow for the creation of custom shelf designs optimized for specific loads and applications.

Safety considerations are paramount in shelf design, particularly in environments where shelves are used to store heavy or valuable items. Proper installation and maintenance are crucial to ensuring long-term stability. Shelves should be securely anchored to walls or other supports, and regular inspections should be conducted to check for signs of wear, deformation, or loosening of supports.

Chapter 25: Airplanes: Lift, Drag, and Flight Principles

Airplanes represent a marvel of human ingenuity and engineering, achieving the seemingly impossible feat of flight by leveraging fundamental principles of physics. Understanding how airplanes fly involves delving into concepts such as lift, drag, thrust, and the overall dynamics of flight. This comprehensive exploration will cover the detailed principles of lift and drag, the forces acting on an aircraft, the role of aerodynamics, the different phases of flight, and the technological advancements that have shaped modern aviation.

At the heart of an airplane's ability to fly is the generation of lift. Lift is the force that opposes gravity and supports the weight of the airplane in the air. It is primarily generated by the wings and is a result of the air pressure differences created above and below the wing surfaces. This phenomenon can be explained using Bernoulli's principle and Newton's third law of motion.

Bernoulli's principle states that an increase in the speed of a fluid occurs simultaneously with a decrease in pressure. When applied to airplane wings, this principle helps explain how lift is generated. An airplane wing is typically designed with a curved upper surface and a flatter lower surface. As the airplane moves forward, air flows over and under the wings. The air traveling over the curved upper surface has to cover a longer distance in the same amount of time as the air flowing under the wing, resulting in faster airflow over the top surface. According to Bernoulli's principle, the faster airflow leads to a decrease in pressure above the wing compared to the pressure below the wing. This pressure difference creates an upward force—lift.

Newton's third law of motion, which states that for every action there is an equal and opposite reaction, also plays a role in lift generation. As the wing deflects air downward, there is an equal and opposite upward force exerted on the wing. This contributes to the lift force that helps keep the airplane aloft.

Drag is another critical force that acts on an airplane, opposing its motion through the air. Drag can be categorized into two main types: parasitic drag and induced drag. Parasitic drag is further divided into form drag, skin friction drag, and interference drag.

Form drag results from the shape of the airplane and the way it displaces air as it moves. It is influenced by the cross-sectional area of the aircraft and the shape of its components. Streamlining the shape of the airplane helps reduce form drag.

Skin friction drag is caused by the friction between the air and the surface of the airplane. The smoother the surface, the less skin friction drag is produced. However, even smooth surfaces have a thin layer of air called the boundary layer that contributes to drag.

Interference drag occurs where different parts of the airplane, such as the wings and fuselage, intersect. The airflow around these intersections is disrupted, creating additional drag.

Induced drag is a byproduct of lift generation. As the wing generates lift, it also creates vortices at the wingtips due to the high-pressure air from beneath the wing spilling over to the low-pressure area above the wing. These vortices create a swirling motion of air that increases drag. Wingtip devices such as winglets are designed to reduce induced drag by minimizing these vortices.

Thrust is the force that propels the airplane forward and is generated by the airplane's engines. The engines work on the principle of Newton's third law of motion, expelling air or exhaust gases backward to produce a forward thrust. In jet engines, this is achieved by compressing air, mixing it with fuel, igniting the mixture, and expelling the high-speed exhaust gases through a nozzle. In propeller-driven airplanes, the engine drives a propeller that accelerates air backward, producing forward thrust.

The balance and interaction of these forces—lift, drag, thrust, and weight—are crucial for stable flight. During level flight, lift must equal the airplane's weight, and thrust must equal drag. When an airplane climbs, the lift must be greater than the weight, and when it descends, the weight must be greater than the lift. Similarly, for acceleration, thrust must exceed drag, and for deceleration, drag must exceed thrust.

Aerodynamics, the study of how air interacts with solid objects, is fundamental to the design and performance of airplanes. The shape of an airplane, known as its aerodynamic profile, is designed to optimize lift and minimize drag. The wings, fuselage, tail, and other components are carefully shaped and positioned to achieve this balance.

The wing's shape, or airfoil, is particularly important. The curvature, or camber, of the wing and the angle at which it meets the oncoming air, known as the angle of attack, play significant roles in lift generation. Increasing the angle of attack increases lift up to a certain point, beyond which the airflow can no longer smoothly follow the wing's contour, leading to a stall where lift suddenly decreases.

Flaps and slats are movable surfaces on the wings that can be extended to increase the wing's surface area and change its shape, allowing for greater lift at lower speeds, which is essential during takeoff and landing.

The tail section of an airplane, which includes the horizontal stabilizer and vertical stabilizer, provides stability and control. The horizontal stabilizer helps maintain pitch control (up and down movement), while the vertical stabilizer aids in yaw control (side-to-side movement). The control surfaces—elevators, ailerons, and rudders—are used by the pilot to maneuver the airplane.

Elevators are hinged surfaces on the horizontal stabilizer that control pitch. Moving the elevators up or down changes the angle of attack of the horizontal stabilizer, causing the airplane's nose to rise or fall.

Ailerons are located on the trailing edges of the wings and control roll (rotation around the longitudinal axis). When one aileron is raised and the other is lowered, they create a differential lift that causes the airplane to roll to one side.

The rudder, located on the vertical stabilizer, controls yaw. Moving the rudder to the left or right causes the airplane's nose to move in the corresponding direction.

The phases of flight—takeoff, climb, cruise, descent, and landing—each present unique challenges and require careful management of the forces acting on the airplane. During takeoff, the engines generate maximum thrust to overcome drag and accelerate the airplane to takeoff speed, at which point the lift generated by the wings exceeds the weight, allowing the airplane to become airborne. The pilot adjusts the angle of attack and uses flaps to optimize lift during this critical phase.

Climb involves gaining altitude while managing the balance between thrust, drag, lift, and weight. The airplane's engines continue to produce high thrust, and the pilot adjusts the pitch to maintain an optimal climb rate without exceeding safe speed limits.

Cruise is the phase where the airplane maintains a steady altitude and speed, typically at high altitude where the thinner air reduces drag and allows for more efficient fuel consumption. The engines produce enough thrust to balance drag, and the airplane is in a state of equilibrium.

Descent involves reducing altitude while controlling speed and managing the forces acting on the airplane. The pilot reduces engine thrust and adjusts the pitch to descend smoothly while maintaining control.

Landing is one of the most challenging phases, requiring precise control to bring the airplane safely to the ground. The pilot uses flaps and other control surfaces to increase lift and drag, allowing for a slower, controlled descent and touchdown. Thrust reversers and brakes are used to decelerate the airplane once it is on the runway.

Technological advancements have significantly enhanced the safety, efficiency, and performance of airplanes. Modern avionics systems provide pilots with critical information about the airplane's status and environment, enabling better decision-making. Autopilot systems can control the airplane during various phases of flight, reducing pilot workload and enhancing safety.

Materials science has also played a crucial role in aviation advancements. The development of lightweight, high-strength materials such as composites and advanced alloys has allowed for stronger, more efficient airframes that improve performance and fuel efficiency.

Aerodynamic innovations, such as winglets and blended wing-body designs, have further reduced drag and improved lift-to-drag ratios, contributing to more efficient flight. Engine technology has also seen significant improvements, with modern jet engines offering greater thrust, fuel efficiency, and reliability compared to earlier designs.

Chapter 26: The Power of Pulleys: Elevating Efficiency

Pulleys are one of the simplest yet most effective tools in the realm of mechanical engineering. Their ability to amplify force and facilitate the movement of heavy loads has made them indispensable in countless applications, from ancient construction projects to modern industrial machinery.

A pulley is essentially a wheel on an axle or shaft that is designed to support the movement of a cable or belt along its circumference. The core principle behind pulleys is the use of a wheel to change the direction of an applied force, making it easier to lift or move heavy objects. The most basic type of pulley, known as a single fixed pulley, changes the direction of the force applied to the rope or cable but does not change the magnitude of the force required to lift the load.

The mechanical advantage of a pulley system is a key concept that explains how pulleys can make lifting tasks easier. Mechanical advantage (MA) is the ratio of the load force to the effort force. For a single fixed pulley, the mechanical advantage is 1, meaning that the effort force required to lift the load is equal to the weight of the load. However, the primary benefit of a single fixed pulley is that it allows the user to apply force in a more convenient direction.

More complex pulley systems, known as block and tackle systems, involve multiple pulleys and provide a greater mechanical advantage. In these systems, pulleys are arranged in sets or "blocks," and the rope is threaded through multiple pulleys. This setup reduces the effort needed to lift a load by distributing the weight across multiple segments of the rope. The mechanical advantage of a block and tackle system is determined by the number of rope segments supporting the load. For example, if a system has four segments of rope supporting the load, the mechanical advantage is 4, meaning the effort force required is only one-fourth of the load's weight.

Pulleys are categorized into several types based on their configuration and purpose. These include fixed pulleys, movable pulleys, and compound pulleys. Fixed pulleys are attached to a fixed point and only change the direction of the force. Movable pulleys, on the other hand, are attached to the load itself and move with it, effectively halving the force needed to lift the load but requiring twice the distance of rope to be pulled. Compound pulleys combine fixed and movable pulleys to maximize mechanical advantage.

The applications of pulleys are vast and varied. They are used in construction, where cranes equipped with pulley systems lift heavy materials to great heights. In elevators, pulleys are crucial for raising and lowering the car efficiently and safely. Pulleys are also found in everyday objects such as window blinds, flagpoles, and exercise equipment, where they help manage and redirect forces to achieve desired outcomes with minimal effort.

Historically, pulleys have played a significant role in human development. The ancient Egyptians are believed to have used basic pulley systems to construct the pyramids, lifting and positioning massive stone blocks with relative ease. The Greeks further refined pulley technology, and the famous Greek mathematician Archimedes is often credited with developing compound pulley systems to move large ships. The Romans adopted and expanded on these technologies, using pulleys in their construction projects and machinery.

In modern times, the principles of pulleys are applied in advanced engineering and machinery. Industrial applications often involve complex pulley systems in conveyor belts, manufacturing equipment, and robotics. These systems are designed to optimize force distribution and efficiency, enabling the smooth and effective operation of heavy machinery and assembly lines.

The efficiency of a pulley system is determined by several factors, including the quality of the materials used, the design of the pulleys and ropes, and the presence of friction. High-quality materials such as strong metals and durable synthetic ropes reduce wear and tear, ensuring longevity and reliable performance. Properly designed pulleys with smooth grooves and low-friction bearings minimize resistance and energy loss, maximizing the system's efficiency.

Friction is an unavoidable factor in any mechanical system, and pulleys are no exception. The friction between the rope and the pulley's surface, as well as within the pulley's bearings, can reduce the overall efficiency of the system. Engineers strive to minimize friction by using lubricants, precision-engineered components, and materials with low friction coefficients. Reducing friction not only enhances efficiency but also extends the lifespan of the pulley system by preventing excessive wear.

In addition to reducing friction, maintaining tension in the rope or belt is crucial for the efficient operation of pulley systems. Proper tension ensures that the rope remains securely within the pulley grooves and transmits force effectively without slipping or sagging. Adjustable tensioning mechanisms are often employed in pulley systems to maintain optimal tension and accommodate changes in load or operating conditions.

Pulleys also play a significant role in safety and ergonomics. By reducing the effort required to lift or move heavy loads, pulleys minimize the risk of injury and strain for workers. This is particularly important in industries where manual handling of heavy objects is common. Ergonomically designed pulley systems help create safer and more efficient working environments, reducing the physical demands on workers and improving overall productivity.

Innovations in pulley technology continue to advance the capabilities and applications of these simple machines. Modern pulley systems incorporate advanced materials, precision engineering, and computer-aided design to achieve higher efficiency, greater load capacity, and enhanced durability. For example, the use of lightweight composite materials and advanced alloys in pulleys and ropes reduces the overall weight of the system while maintaining strength and performance.

Automation and smart technologies are also transforming pulley systems. Automated pulley systems can be controlled and monitored using sensors and software, allowing for precise operation and real-time adjustments. These systems are used in various industries, including manufacturing, logistics, and construction, where they enhance efficiency and safety by providing automated control of lifting and moving operations.

The integration of pulley systems with other mechanical and electrical components has led to the development of sophisticated machines and devices. For instance, in robotics, pulleys are used in conjunction with motors and actuators to create precise and efficient movement. This combination of technologies enables the development of advanced robotic systems capable of performing complex tasks with high accuracy and reliability.

The future of pulley technology holds exciting possibilities, particularly in the fields of automation, robotics, and materials science. Advances in smart materials, such as shape-memory alloys and polymers, could lead to the development of pulleys that automatically adjust their properties in response to changing loads and conditions. This adaptability would further enhance the efficiency and versatility of pulley systems in various applications.

Chapter 27: The Wonder of Water Heaters: Thermodynamics in Action

Water heaters are indispensable appliances in modern households, providing a steady supply of hot water for bathing, cleaning, cooking, and other domestic purposes. The functioning of water heaters is deeply rooted in the principles of thermodynamics, specifically heat transfer and energy conversion.

At the core of a water heater's operation lies the transfer of thermal energy from a heat source to the water, raising its temperature to the desired level. This process involves several thermodynamic principles, including conduction, convection, and radiation.

Conduction is the transfer of heat through direct contact between materials. In a typical water heater, a heat source, such as a gas burner or electric heating element, comes into direct contact with the water, transferring thermal energy to it. The heating element or burner raises the temperature of the surrounding water, creating a temperature gradient that causes heat to flow from the hotter regions to the cooler ones until thermal equilibrium is reached.

Convection plays a crucial role in distributing heat throughout the water tank. As water near the heat source absorbs thermal energy, it becomes less dense and rises to the top of the tank, creating a convection current. Meanwhile, cooler water sinks to the bottom, where it is heated, creating a continuous cycle of circulation that ensures uniform heating of the entire tank.

Radiation is another mechanism by which heat is transferred in water heaters, although it plays a lesser role compared to conduction and convection. Radiation involves the emission of electromagnetic waves, including infrared radiation, which can transfer heat energy to nearby objects without direct contact. In a water heater, the heating element or burner emits infrared radiation that contributes to heating the surrounding water.

Water heaters are available in various types, each employing different mechanisms for heating water. The most common types include storage tank water heaters, tankless (on-demand) water heaters, heat pump water heaters, and solar water heaters.

Storage tank water heaters consist of an insulated tank filled with water, which is heated continuously by a gas burner or electric heating element. As water is used, cold water enters the tank to replace it, and the heating element or burner maintains the water temperature at the desired level.

Tankless water heaters, also known as on-demand water heaters, heat water directly as it passes through a heat exchanger, eliminating the need for a storage tank. When hot water is required, cold water flows through the heat exchanger, where it is rapidly heated to the desired temperature before being delivered to the faucet or shower. Tankless water heaters offer the advantage of providing hot water on demand, without the energy losses associated with storing hot water in a tank.

Heat pump water heaters operate by extracting heat from the surrounding air or ground and transferring it to the water. A heat pump extracts heat from the ambient air or ground using a refrigerant, which is then compressed to increase its temperature before transferring the heat to the water. Heat pump water heaters are highly efficient, as they use ambient heat rather than generating heat directly, making them an environmentally friendly option.

Solar water heaters utilize solar collectors, typically mounted on the roof, to absorb sunlight and convert it into heat. The heat is transferred to a heat exchanger, where it is used to heat the water stored in a tank.

Solar water heaters are renewable and sustainable, relying on free solar energy to provide hot water with minimal environmental impact.

Efficiency is a critical consideration in the design and operation of water heaters, as it directly impacts energy consumption, operating costs, and environmental sustainability. Several factors affect the efficiency of water heaters, including insulation, heating elements, fuel sources, and maintenance.

Proper insulation of the water tank and pipes is essential for minimizing heat losses and improving overall efficiency. A well-insulated tank reduces standby heat losses, which occur when hot water sits unused in the tank and gradually cools over time. Insulating hot water pipes prevents heat from dissipating as the water travels from the heater to the faucet or shower, ensuring that hot water remains hot until it reaches its destination.

The heating elements or burners used in water heaters also play a significant role in efficiency. High-efficiency heating elements and burners are designed to maximize heat transfer to the water while minimizing energy losses. For electric water heaters, efficient heating elements with high thermal conductivity and precise temperature control help reduce energy consumption. For gas water heaters, high-efficiency burners with advanced combustion technology improve fuel efficiency and reduce emissions.

The fuel source used to power water heaters can have a significant impact on efficiency and environmental sustainability. Gas water heaters are commonly fueled by natural gas or propane, which are relatively inexpensive and produce lower greenhouse gas emissions compared to electricity generated from fossil fuels. However, gas water heaters may have lower efficiency ratings compared to electric models, as some energy is lost in the combustion process.

Electric water heaters are powered by electricity and are highly efficient when properly designed and insulated. Electric resistance heating elements convert electrical energy into heat with near-perfect efficiency, making them an attractive option for homeowners with access to renewable energy sources such as solar or wind power.

Regular maintenance is essential for optimizing the efficiency and longevity of water heaters. Routine tasks such as flushing the tank to remove sediment buildup, checking and replacing sacrificial anode rods to prevent corrosion, and inspecting insulation and seals for signs of wear are crucial for ensuring peak performance.

Innovations in water heating technology continue to drive improvements in efficiency, performance, and sustainability. Advanced materials such as high-efficiency insulation and heat exchanger materials enhance heat retention and transfer, reducing energy losses and improving overall efficiency. Smart controls and sensors enable precise temperature regulation and monitoring, allowing users to optimize energy usage and reduce waste. Integration with home automation systems and renewable energy sources further enhances the efficiency and environmental friendliness of water heating systems.

Chapter 28: Gears and Bicycles: The Mechanics of Motion

Gears are fundamental components in the mechanics of motion, playing a crucial role in various mechanical systems, including bicycles. Understanding the principles of gears and their interaction with bicycles offers insights into how these machines efficiently convert human energy into forward motion.

At its core, a gear is a rotating machine part with teeth that mesh with another toothed part to transmit motion and power. Gears are used to change the speed, torque, or direction of motion in mechanical systems. They work by transferring rotational motion from one shaft to another, either increasing or decreasing speed or torque in the process.

In bicycles, gears are primarily used to convert the rotational motion of the rider's legs into forward motion of the wheels. By adjusting the gear ratio, cyclists can adapt to different riding conditions, such as varying terrain, wind resistance, and desired speed. Gearing allows cyclists to maintain an optimal cadence (pedaling rate) and exert the appropriate amount of force to propel the bicycle efficiently.

Bicycles typically use a system of gears known as a drivetrain, which consists of front and rear gears (chainrings and sprockets) connected by a chain. The front gears, located near the pedals, are attached to the crankset, while the rear gears, mounted on the rear wheel hub, are part of the cassette or freewheel.

The gear ratio is a critical parameter that determines the relationship between the number of teeth on the front and rear gears. It represents the ratio of the driven gear's teeth (rear gear) to the driving gear's teeth (front gear) and determines how many revolutions the driven gear completes for each revolution of the driving gear. A higher gear ratio results in faster wheel rotation for a given pedaling cadence, while a lower gear ratio provides greater torque for climbing or accelerating.

The gear ratio affects the mechanical advantage of the drivetrain, influencing the force required to pedal and the speed at which the bicycle travels. A higher gear ratio (larger front gear or smaller rear gear) requires more force to pedal but results in higher speeds on flat terrain or downhill. Conversely, a lower gear ratio (smaller front gear or larger rear gear) reduces the force required to pedal but allows for easier climbing or riding against headwinds.

The selection of gear ratios depends on various factors, including the cyclist's fitness level, riding style, terrain, and weather conditions. Experienced cyclists learn to anticipate changes in terrain and adjust their gearing accordingly to maintain an optimal cadence and efficiency. In competitive cycling, gear selection can be critical for gaining a competitive edge, particularly in races with variable terrain or challenging conditions.

Bicycles can feature different types of gear systems, each offering unique advantages and characteristics. The most common types include single-speed, multi-speed (derailleur), and internally geared systems.

Single-speed bicycles have only one gear ratio and are popular for their simplicity, low maintenance, and reliability. They are often favored for urban commuting, casual riding, and fixed-gear cycling disciplines.

Multi-speed bicycles utilize derailleurs, which are mechanisms that move the chain between different-sized gears to change the gear ratio. Derailleur systems offer a wide range of gear ratios, allowing cyclists to tackle various riding conditions with ease. They are commonly found on road bikes, mountain bikes, and hybrid bicycles.

Internally geared hubs are another type of gear system that houses all the gears within the rear hub of the bicycle. Unlike derailleur systems, internally geared hubs provide gear shifting without external components, making them more resistant to damage from impacts, dirt, and weather. They offer a clean and low-maintenance option for cyclists seeking versatility and reliability.

Efficiency is a key consideration in gear design and selection, particularly in bicycles where energy conservation is paramount. Gear efficiency refers to the ratio of output power to input power, with higher efficiency indicating less energy loss in the gear system. Factors such as friction, alignment, material properties, and lubrication affect gear efficiency and can impact cycling performance.

To maximize efficiency, gear systems are designed to minimize friction and wear between moving parts. Precision machining, high-quality materials, and proper lubrication help reduce energy losses and prolong the lifespan of gears and drivetrain components. Regular maintenance, including cleaning, lubrication, and adjustment, is essential for ensuring smooth and efficient gear operation.

Innovations in gear technology continue to drive advancements in cycling performance and efficiency. Lightweight materials, such as carbon fiber and titanium, offer strength and durability while minimizing weight, enhancing the responsiveness and agility of bicycles. Aerodynamic designs optimize airflow around gears and drivetrain components, reducing drag and improving overall efficiency.

Electronic shifting systems, which use motors and sensors to control gear changes electronically, offer precise and instantaneous shifting with minimal effort. These systems provide seamless gear transitions and customizable shifting patterns, allowing cyclists to fine-tune their gearing for maximum performance and comfort.

The future of gear technology holds promise for further improvements in efficiency and performance. Ongoing research focuses on developing advanced materials with enhanced strength-to-weight ratios and friction-reducing properties to further improve gear efficiency. Computational modeling and simulation techniques enable engineers to optimize gear designs for specific applications, taking into account factors such as load distribution, stress concentrations, and vibration damping.

Additionally, advancements in manufacturing processes, such as additive manufacturing (3D printing), allow for the creation of complex gear geometries with high precision and customization. This flexibility opens up new possibilities for designing gears tailored to the unique requirements of individual cyclists, optimizing performance while reducing weight and energy consumption.

Integration with smart technologies, such as sensors and wireless connectivity, holds potential for enhancing gear systems with real-time performance monitoring and adaptive control. By collecting data on cycling metrics, environmental conditions, and rider preferences, smart gear systems can automatically adjust gearing parameters to optimize efficiency, comfort, and safety.

Chapter 29: The Physics of Fishing: Casting and Reeling

The physics of fishing encompasses a fascinating interplay of mechanical principles, fluid dynamics, and human biomechanics, all working together to facilitate the art of angling. From the precise mechanics of casting a fishing line to the dynamics of reeling in a catch, understanding the physics behind fishing techniques enhances both the enjoyment and success of this popular recreational activity.

Casting a fishing line involves a complex series of motions that rely on fundamental principles of physics, including Newton's laws of motion, rotational dynamics, and fluid mechanics. The primary goal of casting is to propel the fishing lure or bait to a desired location with accuracy and precision. Achieving this requires careful coordination of the angler's movements, the rod's action, and the properties of the fishing line.

Newton's laws of motion are fundamental to understanding the mechanics of casting. According to Newton's first law, an object at rest will remain at rest unless acted upon by an external force, while an object in motion will remain in motion unless acted upon by an external force. When casting a fishing line, the angler applies a force to the rod, causing it to bend and store potential energy in the form of elastic deformation. As the rod is released, this stored energy is transferred to the fishing line, propelling the lure forward.

Newton's second law relates the acceleration of an object to the force applied to it and its mass. In the context of casting, the force exerted by the angler's hand and arm, combined with the flexibility and action of the fishing rod, determines the speed and distance at which the lure is cast. By increasing the force applied or reducing the mass of the lure, anglers can achieve longer and more powerful casts.

Rotational dynamics govern the motion of the fishing rod during casting, particularly the rotation of the rod tip as it accelerates and decelerates. The angular velocity and acceleration of the rod tip contribute to the speed and trajectory of the fishing line, influencing the accuracy and distance of the cast. Anglers often use techniques such as "loading" the rod by bending it backward before casting and "unloading" it with a quick, smooth motion to maximize casting distance and control.

Fluid mechanics plays a crucial role in the behavior of the fishing line and lure as they move through the air and water. The aerodynamics of the fishing line, lure shape, and air resistance affect the trajectory and accuracy of the cast. Factors such as wind speed and direction further influence the flight of the lure, requiring anglers to adjust their casting technique accordingly.

Once the lure lands on the water, fluid dynamics come into play as it interacts with the water surface and underlying currents. The buoyancy and hydrodynamics of the lure determine its movement and presentation, mimicking the behavior of natural prey and attracting the attention of fish. Anglers may employ techniques such as "walking the dog" or "jigging" to impart lifelike movement to the lure and entice strikes from fish.

Reeling in a fish involves overcoming various forces and resistances, including friction between the fishing line and guides, water resistance, and the fish's own resistance. The drag force, generated by the fish pulling against the fishing line, must be balanced by the angler's applied force to prevent the line from breaking or the fish from escaping.

The mechanics of reeling rely on principles of biomechanics, particularly the use of leverage and muscle coordination to exert force and control over the fishing rod. By leveraging the mechanical advantage provided by the reel handle and rod, anglers can apply greater force to reel in larger fish with less effort. Proper body mechanics, including stance, posture, and grip, are essential for maintaining balance and control while fighting a fish.

Understanding the physics of fishing techniques enables anglers to optimize their performance, adapt to changing conditions, and increase their chances of success on the water. By applying principles such as Newton's laws, rotational dynamics, and fluid mechanics, anglers can cast with greater accuracy, present lures more effectively, and land more fish. Whether pursuing freshwater bass, saltwater marlin, or fly-fishing trout, a solid grasp of the physics behind fishing enhances the experience and appreciation of this timeless pursuit.

Chapter 30: Magnets: Attraction, Repulsion, and Practical Uses

Magnets are fascinating objects that have captured human curiosity and imagination for centuries. Their ability to attract and repel other magnets and certain materials is rooted in the fundamental principles of electromagnetism and quantum mechanics. Beyond their intrinsic allure, magnets have a myriad of practical uses in everyday life, from household appliances to advanced technological applications.

At its core, a magnet is an object that produces a magnetic field, which exerts attractive or repulsive forces on other magnets or magnetic materials. This magnetic field is generated by the alignment of magnetic domains within the material, where the magnetic moments of individual atoms or molecules align in a coordinated manner. This alignment creates a net magnetic dipole moment, resulting in the manifestation of magnetic properties.

The behavior of magnets is governed by Maxwell's equations of electromagnetism, which describe the relationship between electric fields, magnetic fields, and electric charges. According to Maxwell's equations, changing electric fields can induce magnetic fields, and vice versa, through electromagnetic induction. This phenomenon forms the basis of electromagnetism and is central to understanding the underlying principles of magnetism.

One of the most remarkable properties of magnets is their ability to attract certain materials while repelling others. This behavior is governed by the orientation of magnetic dipoles within the material and the interaction of their magnetic fields. Like poles repel each other, while opposite poles attract, following the basic principle of magnetic attraction and repulsion.

The magnetic field surrounding a magnet extends into the surrounding space, creating regions of influence where the magnetic force can be felt. These regions, known as magnetic fields or flux fields, are characterized by field lines that emanate from the magnet's poles and loop back around to form closed paths. The density and orientation of these field lines indicate the strength and direction of the magnetic field, providing insights into the magnet's properties and behavior.

The strength of a magnet's magnetic field is quantified by its magnetic flux density, measured in units of teslas (T) or gauss (G). The magnetic flux density depends on factors such as the material composition, shape, and size of the magnet, as well as external influences such as temperature and applied magnetic fields.

Permanent magnets, such as those made from iron, nickel, cobalt, and certain rare-earth metals, retain their magnetic properties even in the absence of an external magnetic field. These materials have a high magnetic permeability, allowing them to maintain strong magnetic fields and retain their magnetization over time.

In contrast, temporary magnets, also known as electromagnets, rely on the flow of electric current to generate a magnetic field. By passing an electric current through a coil of wire wound around a ferromagnetic core, such as iron or steel, temporary magnets can be created with controllable magnetic properties. Electromagnets are widely used in various applications, including electric motors, transformers, and magnetic resonance imaging (MRI) machines.

The practical applications of magnets span a wide range of industries and technologies, demonstrating their versatility and importance in modern society. In electronics and telecommunications, magnets are used in speakers, headphones, microphones, and magnetic storage devices such as hard drives and magnetic tape.

In the medical field, magnets play a crucial role in diagnostic imaging techniques such as MRI, where powerful magnets generate magnetic fields to create detailed images of the body's internal structures. Magnets are also used in magnetic therapy for pain relief and in magnetic resonance spectroscopy for chemical analysis.

In transportation, magnets are utilized in electric motors, generators, and magnetic levitation (maglev) trains, where they enable efficient energy conversion and propulsion. Maglev trains, in particular, use powerful electromagnets to levitate above the tracks and propel the train forward at high speeds, offering a quiet, smooth, and energy-efficient mode of transportation.

In manufacturing and engineering, magnets are used in a variety of applications, including magnetic separation, material handling, and lifting and clamping systems. Magnetic separators are employed to remove unwanted ferrous contaminants from raw materials, while magnetic lifting systems enable the safe and efficient handling of heavy objects in industrial settings.

In renewable energy technologies, magnets are essential components in wind turbines, electric vehicle motors, and hydroelectric generators, where they contribute to the conversion of mechanical energy into electrical energy. The use of magnets in these applications helps reduce reliance on fossil fuels and mitigate environmental impacts associated with traditional energy sources.

The development of new materials and technologies, such as rare-earth magnets and magnetic nanoparticles, continues to expand the capabilities and applications of magnets in various fields. Rare-earth magnets, composed of alloys containing rare-earth elements such as neodymium and samarium, exhibit exceptional magnetic properties and are used in high-performance applications such as aerospace, defense, and renewable energy.

Magnetic nanoparticles, on the other hand, are microscopic particles with magnetic properties that can be manipulated and controlled for various biomedical and environmental applications. These nanoparticles are used in drug delivery systems, magnetic resonance imaging (MRI) contrast agents, and environmental remediation techniques, demonstrating the potential of magnets in addressing complex challenges in healthcare and environmental science.

Chapter 31: The Mechanics of Makeup: The Science of Cosmetics

The mechanics of makeup delve into the intricate science behind cosmetics, exploring the formulation, application, and interactions of various products with the skin. Makeup has been used for centuries as a form of self-expression, cultural tradition, and enhancement of natural beauty. Understanding the underlying principles of cosmetics involves a deep dive into chemistry, materials science, dermatology, and even psychology.

Formulating cosmetics involves blending a variety of ingredients to achieve desired properties such as color, texture, coverage, and longevity. These ingredients can be categorized into several broad classes, including pigments, binders, emollients, thickeners, preservatives, and additives. Each ingredient plays a specific role in the formulation and performance of the product.

Pigments are responsible for imparting color to makeup products, ranging from foundation and blush to eyeshadow and lipstick. These pigments can be organic or inorganic compounds, each with its own unique color characteristics and stability properties. Binders help ingredients adhere to the skin and improve the longevity of the product, while emollients provide hydration and smoothness, enhancing the texture and feel of the makeup.

Thickeners and stabilizers are used to adjust the viscosity and consistency of cosmetics, ensuring easy application and uniform coverage. Preservatives are essential for preventing microbial growth and prolonging the shelf life of products, while additives such as antioxidants and UV filters provide additional benefits such as protection against environmental damage.

The application of makeup involves a combination of artistry, technique, and knowledge of facial anatomy. Different makeup products require specific application techniques to achieve desired effects, whether it's creating a flawless complexion, accentuating facial features, or experimenting with bold, creative looks. Brushes, sponges, and fingers are commonly used tools for applying makeup, each offering unique advantages in terms of precision, blending, and coverage.

Foundation serves as the base for makeup, evening out skin tone and providing a smooth canvas for further application. Concealers are used to cover imperfections such as blemishes, dark circles, and discolorations, while powders help set makeup in place and control shine. Blush and bronzer add color and dimension to the face, while eyeshadow, eyeliner, and mascara enhance the eyes and create different effects.

Lipstick, lip gloss, and lip liner are used to enhance the lips, providing color, shine, and definition. Each type of makeup product comes in a variety of formulations, finishes, and shades, allowing individuals to customize their look according to personal preferences and occasions.

The physiological effects of makeup on the skin are a topic of interest in dermatology and cosmetic science. While makeup can enhance appearance and boost confidence, improper use or low-quality products can lead to adverse reactions such as irritation, allergies, and acne. Understanding skin types, sensitivities, and compatibility with makeup ingredients is crucial for selecting products that are safe and suitable for individual needs.

Advancements in cosmetic science and technology have led to the development of innovative formulations and products designed to address specific skin concerns and preferences. From mineral makeup and water-resistant formulations to long-wear and multi-functional products, cosmetics continue to evolve to meet the diverse needs of consumers.

The psychology of makeup explores the emotional and psychological effects of cosmetics on self-perception, confidence, and social interactions. Studies have shown that wearing makeup can influence perceptions of attractiveness, competence, and trustworthiness, affecting interpersonal relationships and professional success. Makeup can also serve as a form of self-expression, empowerment, and creativity, allowing individuals to experiment with different looks and identities.

Chapter 32: Solar Panels: Harnessing the Power of the Sun

Solar panels are revolutionary devices that harness the abundant energy of the sun to generate electricity through the photovoltaic effect. This transformative technology has emerged as a cornerstone of renewable energy, offering a clean, sustainable alternative to traditional fossil fuel-based power generation. Understanding the mechanics, components, and applications of solar panels involves delving into the principles of solar energy conversion, materials science, engineering, and environmental sustainability.

At the heart of solar panels lies the photovoltaic effect, a phenomenon first discovered in the 19th century that describes the generation of electric current when certain materials are exposed to light. Solar panels consist of interconnected solar cells, typically made from semiconductor materials such as silicon, that convert sunlight directly into electricity through this process. When photons from sunlight strike the surface of a solar cell, they transfer their energy to electrons in the material, causing them to become excited and create an electric current.

The efficiency of solar panels in converting sunlight into electricity depends on various factors, including the type of solar cell, the quality of the materials used, and environmental conditions such as sunlight intensity and temperature. Advances in photovoltaic technology have led to the development of increasingly efficient solar cells, with efficiencies exceeding 20% in some cases.

Solar cells are typically arranged in a grid-like pattern on a solar panel, with multiple panels connected together to form a solar array. The electrical output of the solar panels is then combined and routed through an inverter, which converts the direct current (DC) generated by the panels into alternating current (AC) suitable for use in homes, businesses, and the electrical grid.

The construction of solar cells involves several layers of semiconductor materials that work together to facilitate the photovoltaic process. The top layer, known as the anti-reflective coating, helps maximize sunlight absorption by reducing reflection and increasing light transmission into the cell. Beneath the anti-reflective coating lies the semiconductor material, typically silicon in crystalline or thin-film form, which absorbs photons and generates electric current.

Below the semiconductor layer are metal contacts that collect the generated electricity and conduct it out of the solar cell. These contacts are typically made from materials such as silver or aluminum and are carefully designed to minimize resistance and maximize electrical conductivity. Finally, a protective encapsulation layer seals the solar cell against environmental factors such as moisture, dust, and mechanical damage.

Solar panels are versatile and adaptable, with applications ranging from residential and commercial rooftop installations to utility-scale solar farms and off-grid power systems. Rooftop solar installations allow homeowners and businesses to generate their own electricity, reducing reliance on grid-supplied power and lowering energy bills. Solar farms, consisting of large arrays of solar panels, can generate megawatts of electricity and feed it directly into the electrical grid, contributing to the transition to clean, renewable energy sources.

Installation considerations for solar panels include factors such as roof orientation, tilt angle, shading, and local climate conditions. Ideally, solar panels should be installed on south-facing roofs with minimal shading and at an angle optimized for maximum sunlight exposure throughout the year. Mounting systems such as racks or frames are used to secure the solar panels to the roof or ground and ensure proper alignment and stability.

In addition to their environmental benefits, solar panels offer significant economic advantages, including long-term cost savings, energy independence, and potential revenue generation through net metering and renewable energy incentives. While the upfront cost of solar panel installation can be substantial, it is often offset by savings on electricity bills and the availability of financial incentives such as tax credits, rebates, and feed-in tariffs.

The environmental impact of solar panels is relatively low compared to conventional fossil fuel-based power generation. Solar energy is abundant, renewable, and emits no greenhouse gases or air pollutants during operation. However, the production and disposal of solar panels do have environmental consequences, including energy consumption, resource depletion, and waste generation. Efforts to improve the sustainability of solar panel manufacturing and recycling processes are ongoing, with advancements in materials recycling, resource recovery, and eco-friendly manufacturing techniques.

Chapter 33: The Physics of Baking: Heat Transfer in Ovens

The physics of baking encompasses a fascinating array of scientific principles, particularly the intricate dynamics of heat transfer within ovens. Baking, a culinary art form dating back millennia, involves the transformation of raw ingredients into a wide variety of delicious treats through the application of heat. Understanding the physics behind baking not only enhances culinary skills but also sheds light on fundamental principles of thermodynamics, conduction, convection, and radiation.

At its core, baking relies on the controlled application of heat to transform raw ingredients such as flour, sugar, eggs, and leavening agents into finished products such as bread, cakes, cookies, and pastries. The process of baking involves several key steps, including mixing, shaping, proofing (for yeast-leavened dough), and finally, baking in an oven.

The physics of baking begins with the preheating of the oven, where thermal energy is transferred from the heating elements or gas burners to the interior of the oven chamber. Preheating ensures that the oven reaches the desired temperature before placing the food inside, promoting even cooking and consistent results.

Once the oven is preheated, heat transfer mechanisms come into play to cook the food. Conduction, convection, and radiation are the primary methods of heat transfer involved in baking, each contributing to the overall cooking process in different ways.

Conduction occurs when heat is transferred directly from the oven walls, racks, or baking sheets to the food through direct contact. In baking, conduction plays a crucial role in ensuring that the bottom of baked goods receives sufficient heat to cook evenly and develop a golden-brown crust. Baking sheets and pans made from materials with high thermal conductivity, such as aluminum or steel, facilitate efficient conduction and promote even browning.

Convection involves the transfer of heat through the movement of air or steam within the oven chamber. In conventional ovens, air is heated by the heating elements or gas burners and then circulated throughout the oven by fans or natural convection currents. This circulating air helps distribute heat evenly around the food, speeding up the cooking process and ensuring uniform doneness. Convection ovens, which feature dedicated fans for air circulation, are particularly effective for baking foods that require rapid and even cooking, such as pastries and delicate baked goods.

Radiation refers to the transfer of heat through electromagnetic waves, such as infrared radiation, emitted by the heating elements or gas burners in the oven. These radiant heat waves penetrate the surface of the food and transfer thermal energy to the interior, promoting browning and caramelization while also helping to cook the food from the inside out. Radiant heat is particularly important for achieving crispy crusts on bread and pastries and for caramelizing sugars in baked goods.

The temperature of the oven plays a critical role in the baking process, influencing the rate of chemical reactions, moisture evaporation, and heat transfer within the food. Different types of baked goods require specific oven temperatures to achieve optimal results. For example, bread typically bakes best at higher temperatures (around 400-450°F or 200-230°C) to promote oven spring and crust development, while delicate pastries and cakes may require lower temperatures (around 325-375°F or 160-190°C) to prevent over-browning and ensure a tender crumb.

In addition to heat transfer mechanisms, other factors such as oven humidity, baking time, and the presence of steam can also affect the outcome of baked goods. Steam, for example, is often used in baking to create a moist environment that promotes oven spring, enhances crust formation, and prevents excessive drying of the food. Techniques such as steam injection or placing a pan of water in the oven can help achieve these effects.

Furthermore, the physical and chemical transformations that occur during baking are governed by principles of food chemistry and biochemistry. Ingredients such as flour, sugar, and eggs undergo complex reactions such as protein denaturation, starch gelatinization, and Maillard browning, resulting in the formation of desirable flavors, textures, and colors in the finished baked goods.

Chapter 34: Fireplaces: Combustion and Heat Distribution

Fireplaces are iconic fixtures in homes around the world, offering warmth, ambiance, and a focal point for gatherings. Understanding the intricate dynamics of combustion and heat distribution within fireplaces involves delving into the principles of thermodynamics, fluid mechanics, and heat transfer. From the initial ignition of fuel to the circulation of warm air throughout a room, fireplaces represent a fascinating intersection of science, engineering, and architectural design.

At the heart of every fireplace lies the process of combustion, where fuel is oxidized in the presence of oxygen to release heat and light. The type of fuel used in a fireplace can vary widely, ranging from wood and natural gas to pellets and biofuels. Regardless of the fuel source, the combustion process follows the same basic principles, involving three essential components: fuel, oxygen, and heat.

When fuel, such as wood logs, is ignited in a fireplace, it undergoes a series of chemical reactions known as combustion. In the presence of heat from the initial ignition, the fuel releases volatile gases such as methane, carbon monoxide, and hydrogen, which then mix with oxygen from the surrounding air. These gases react with oxygen to produce heat, water vapor, carbon dioxide, and other byproducts, releasing energy in the form of heat and light.

The efficiency of combustion in a fireplace depends on various factors, including the moisture content and type of fuel, the availability of oxygen, and the design of the fireplace itself. Proper airflow is essential for ensuring complete combustion and minimizing the production of pollutants such as carbon monoxide and particulate matter. Insufficient airflow can result in inefficient combustion, leading to the buildup of creosote and soot in the chimney, reduced heat output, and increased air pollution.

Fireplaces employ various mechanisms to control airflow and optimize combustion efficiency. Traditional masonry fireplaces rely on natural convection currents to draw in air from the room and exhaust combustion gases through the chimney. Metal fireplaces and inserts may feature adjustable air vents or dampers that allow users to regulate airflow and combustion intensity manually. Modern high-efficiency fireplaces may incorporate advanced combustion technologies such as secondary air injection, catalytic converters, or gasification chambers to achieve cleaner, more efficient combustion.

In addition to producing radiant heat directly from the flames, fireplaces also transfer heat to the surrounding environment through convection and radiation. Convection occurs as warm air rises from the fireplace and circulates throughout the room, creating a natural convection current that distributes heat evenly. Radiant heat, meanwhile, emanates from the hot surfaces of the fireplace, including the firebox, hearth, and surrounding masonry or metal components, warming nearby objects and occupants.

The design and construction of the fireplace play a crucial role in determining its heating efficiency and effectiveness. Factors such as the size and shape of the firebox, the materials used in construction, and the presence of heat-reflective surfaces can all influence the fireplace's heat output and distribution. Masonry fireplaces, with their thick brick or stone walls, tend to retain heat more effectively than metal fireplaces, providing longer-lasting warmth even after the fire has died down.

Fireplace inserts, which are prefabricated metal units installed inside existing fireplaces, offer improved efficiency and heat output by capturing and circulating heat more effectively. Inserts typically feature insulated fireboxes, adjustable air vents, and built-in blowers or fans that help distribute warm air throughout the room. Some inserts also include heat exchangers or catalytic converters to extract additional heat from combustion gases before they are vented outside.

Chapter 35: The Power of Pencils: Graphite and Friction

The power of pencils lies in their humble yet remarkable composition, particularly the use of graphite and the interplay of friction between the pencil lead and writing surface. Pencils have been essential writing tools for centuries, offering a portable, versatile, and inexpensive means of expression for artists, writers, students, and professionals alike. Understanding the physics behind pencils involves exploring the properties of graphite, the mechanics of friction, and the impact of pencil design and usage on writing performance.

Graphite, the primary component of pencil lead, is a form of carbon known for its unique structure and properties. Unlike traditional writing materials such as ink or charcoal, graphite consists of layers of hexagonally arranged carbon atoms bonded together in a two-dimensional sheet. These layers are held together by weak van der Waals forces, allowing them to slide over each other easily and giving graphite its distinctive slippery feel.

The structure of graphite gives rise to its exceptional lubricating properties, making it an ideal material for writing instruments such as pencils. When a pencil is drawn across a surface, the friction between the graphite and the paper causes the layers of graphite to shear off and deposit onto the surface, leaving behind a mark. This process, known as abrasion or wear, is influenced by factors such as the pressure applied, the angle of the pencil, and the roughness of the writing surface.

The mechanics of friction play a crucial role in determining the quality and characteristics of the mark left by a pencil. Friction is the force that opposes the relative motion or tendency of motion between two surfaces in contact. In the case of pencils, friction between the graphite and the paper surface creates the resistance necessary to transfer graphite particles onto the paper, forming visible lines or marks.

The coefficient of friction, a measure of the resistance to sliding between two surfaces, depends on factors such as surface roughness, material properties, and the presence of lubricants or additives. Graphite's low coefficient of friction allows it to glide smoothly over paper surfaces, creating crisp, uniform lines with minimal effort.

The hardness of a pencil, determined by the proportion of clay and graphite in the pencil lead, also influences its frictional properties and writing performance. Pencils are typically graded according to their hardness, with higher grades (such as 2H or 4H) containing more clay and producing lighter, finer lines, while lower grades (such as 2B or 4B) contain more graphite and produce darker, bolder lines.

Harder pencils require more pressure to produce visible marks and may feel scratchy or uncomfortable to use on rough paper surfaces. Softer pencils, on the other hand, require less pressure and produce smoother, more expressive lines, making them preferred for sketching, shading, and artistic applications.

Surface roughness also plays a significant role in the frictional interaction between pencils and paper. Smoother paper surfaces provide less resistance to pencil movement and tend to produce cleaner, more consistent lines, while rougher surfaces may cause pencils to catch or skip, resulting in uneven or jagged marks.

In addition to their frictional properties, pencils are also influenced by factors such as lead diameter, point sharpness, and eraser quality, all of which can affect writing comfort, precision, and erasability. Thicker leads and sharper points provide greater control and detail, while softer erasers and better eraser attachments improve the ability to correct mistakes and modify drawings.

Chapter 36: Escalators: Continuous Motion Mechanisms

Escalators are ubiquitous fixtures in modern urban environments, providing efficient vertical transportation in places such as airports, train stations, shopping malls, and office buildings. Understanding the intricate mechanisms behind escalators involves delving into the principles of mechanical engineering, dynamics, and control systems. From their robust structural design to the sophisticated motor-driven components that facilitate continuous motion, escalators represent a marvel of engineering ingenuity and innovation.

At the heart of every escalator lies a series of interconnected steps or treads that travel along a looped track in a continuous cycle. The steps are mounted on a pair of linked chains or belts that are driven by motorized sprockets or pulleys located at both ends of the escalator. As the steps move along the track, passengers can step on or off the escalator at designated entry and exit points, allowing for smooth and uninterrupted vertical transportation.

The structural components of an escalator are designed to withstand the stresses and loads associated with continuous operation and passenger traffic. The main structural elements include the truss or framework that supports the escalator, the track assembly that guides the movement of the steps, and the handrails that provide stability and support for passengers.

The drive system of an escalator consists of an electric motor, gearbox, and drive chains or belts that transmit power from the motor to the sprockets or pulleys. The motor is typically located in a machinery room beneath the escalator or housed within the escalator structure itself. The gearbox serves to increase the torque and reduce the speed of the motor output, allowing for smooth and controlled operation of the escalator.

The drive chains or belts are connected to the motor-driven sprockets or pulleys at both ends of the escalator, forming a closed loop that drives the movement of the steps. The speed and direction of the escalator can be controlled electronically using variable-frequency drives (VFDs) or other speed control mechanisms, allowing for adjustments to match passenger flow and operational requirements.

Safety features are an essential aspect of escalator design and operation, ensuring the well-being of passengers and preventing accidents or injuries. Common safety features include emergency stop buttons, handrail sensors, and safety skirts or brushes that prevent objects from becoming trapped between the steps and the skirt panels. In addition, escalators are equipped with sensors and detectors that monitor speed, alignment, and other parameters to detect and respond to abnormal conditions or malfunctions.

The physics of motion governing escalators involves principles of mechanics, dynamics, and friction. As the steps move along the track, they undergo a series of mechanical motions, including rotation, translation, and acceleration. Friction between the steps and the track surface provides traction and prevents slipping, allowing passengers to step on and off the escalator safely.

The efficiency and performance of escalators are influenced by factors such as load capacity, speed, and duty cycle. Escalators are designed to accommodate varying passenger loads and traffic patterns, with larger escalators featuring wider steps and higher capacities to handle peak demand. Speed is another critical factor, with escalators typically operating at speeds ranging from 0.5 to 1.0 meters per second (1.6 to 3.3 feet per second) to ensure efficient passenger flow without compromising safety.

Maintenance and upkeep are essential aspects of escalator operation, ensuring optimal performance, reliability, and safety. Regular inspections, lubrication, and adjustments are necessary to keep escalators in good working condition and prevent premature wear and tear. Escalator technicians and maintenance personnel undergo specialized training to diagnose and address issues such as mechanical failures, electrical faults, and safety hazards.

Chapter 37: The Resilience of Rubber Bands: Elasticity and Hooke's Law

Rubber bands, despite their simple appearance, exhibit a fascinating array of properties that make them indispensable in various applications, from office supplies to industrial machinery. Central to their functionality is the concept of elasticity and the principles outlined by Hooke's Law, which describe the behavior of elastic materials under stress and deformation. Understanding the resilience of rubber bands involves exploring the physics of elasticity, the molecular structure of rubber, and the practical applications of rubber bands in everyday life.

Rubber bands are made from natural rubber or synthetic elastomers, which are polymers with unique molecular structures that impart elasticity and flexibility to the material. In natural rubber, long chains of hydrocarbon molecules are cross-linked by sulfur atoms to form a three-dimensional network known as a vulcanized rubber. This network structure allows rubber to stretch and deform under applied stress and return to its original shape when the stress is removed, making it ideal for use in elastic products such as rubber bands.

The elasticity of rubber bands arises from the reversible stretching and recoiling of polymer chains within the material in response to applied forces. When a rubber band is stretched, the polymer chains are pulled apart, causing them to straighten and align in the direction of the applied force. As the stretching force is increased, more polymer chains become extended, resulting in greater deformation of the material.

Hooke's Law describes the relationship between the force applied to an elastic material and the resulting deformation or extension of the material. According to Hooke's Law, the force required to stretch or compress an elastic material is directly proportional to the amount of deformation produced, as long as the material remains within its elastic limit. Mathematically, this relationship is expressed as $F = kx$, where F is the force applied, k is the stiffness or spring constant of the material, and x is the displacement or extension of the material from its equilibrium position.

In the case of rubber bands, Hooke's Law provides a useful framework for understanding their stress-strain behavior and predicting their response to applied forces. When a rubber band is stretched, the force exerted by the band is proportional to the amount of stretch or extension produced, allowing for precise control over the tension and elasticity of the band. This property makes rubber bands versatile tools for a wide range of applications, from securing objects together to providing tension in mechanical systems.

The resilience of rubber bands is further enhanced by their ability to undergo large deformations without permanent damage or failure. This property, known as extensibility, allows rubber bands to stretch to several times their original length and return to their original shape repeatedly without losing their elasticity. The high tensile strength of rubber bands also enables them to withstand considerable forces without breaking or tearing, making them durable and reliable in demanding environments.

Rubber bands find countless practical applications in everyday life, from bundling and organizing items to providing tension in mechanical systems and experiments. In offices, schools, and homes, rubber bands are commonly used to secure stacks of paper, seal bags and containers, and create makeshift tools and devices. In industrial settings, rubber bands serve as essential components in machinery and equipment, providing tension, damping vibrations, and transmitting power in various applications.

In addition to their practical utility, rubber bands also serve as educational tools for exploring concepts such as elasticity, stress, and strain in physics and engineering classrooms. Through hands-on experiments and demonstrations, students can investigate the behavior of rubber bands under different loading conditions, analyze stress-strain curves, and apply mathematical models such as Hooke's Law to predict and understand their mechanical behavior.

Chapter 38: Headphones: Sound Isolation and Acoustic Engineering

Headphones have revolutionized the way we experience audio, providing a personal and immersive listening environment in a wide range of settings. Central to their design is the concept of sound isolation and the principles of acoustic engineering, which govern how headphones reproduce sound while minimizing external noise. Understanding the intricacies of headphones involves delving into topics such as transducer technology, earcup design, noise cancellation techniques, and the psychoacoustic principles that shape our perception of sound.

At the heart of every pair of headphones is a transducer, which converts electrical signals into sound waves that we perceive as audio. There are two main types of transducers used in headphones: dynamic drivers and balanced armature drivers. Dynamic drivers are the most common type and consist of a diaphragm attached to a coil of wire suspended in a magnetic field. When an electrical signal is applied to the coil, it creates a magnetic field that interacts with the magnetic field of the driver, causing the diaphragm to vibrate and produce sound waves. Balanced armature drivers, on the other hand, use a tiny armature suspended between two magnets to generate sound waves when an electrical current is passed through it. These drivers are typically smaller and lighter than dynamic drivers, making them ideal for in-ear and earbud-style headphones.

The design of the earcups or earpieces plays a crucial role in determining how effectively headphones isolate sound from the surrounding environment. Closed-back headphones feature earcups that fully enclose the ears, forming a sealed acoustic chamber that blocks out external noise and prevents sound leakage. This design provides excellent passive noise isolation and enhances the listening experience, particularly in noisy or crowded environments. Open-back headphones, by contrast, have earcups with perforations or vents that allow air and sound to pass through more freely. While open-back headphones often provide a more natural and spacious soundstage, they offer less isolation from external noise and may disturb others nearby due to sound leakage.

In addition to passive noise isolation, many headphones employ active noise cancellation (ANC) technology to further reduce unwanted external noise. ANC headphones feature built-in microphones that capture ambient sound and generate anti-noise signals to cancel out incoming noise waves. These anti-noise signals are then mixed with the audio signal from the source, effectively canceling out external noise and creating a quieter listening environment. ANC headphones are particularly effective at reducing low-frequency sounds such as engine rumble and background chatter, making them popular choices for travel and commuting.

The effectiveness of sound isolation and noise cancellation in headphones depends on various factors, including the design of the earcups, the quality of the transducers, and the efficiency of the ANC circuitry. Proper fit and seal are also critical for maximizing sound isolation, as gaps between the earcups and the ears can allow external noise to leak in and compromise the listening experience. Many headphones come with adjustable headbands, swiveling earcups, and interchangeable earpads to ensure a comfortable and secure fit for users of all shapes and sizes.

Psychoacoustic principles also play a significant role in headphone design and engineering, influencing how we perceive sound quality, spatial imaging, and frequency response. Factors such as frequency response, soundstage, and transient response contribute to the overall sonic character of headphones and determine how accurately they reproduce recorded audio. Engineers carefully tune the acoustic properties of headphones to optimize clarity, balance, and immersion, taking into account psychoacoustic phenomena such as the masking effect, binaural cues, and auditory localization.

Chapter 39: How Zippers Work: Interlocking Mechanisms

Zippers are ubiquitous fastening devices found in clothing, bags, and countless other everyday items, offering a convenient and reliable way to open and close fabric openings. Central to their functionality is a complex yet elegant interlocking mechanism that allows for secure fastening while remaining easy to operate. Understanding how zippers work involves exploring the intricate design of their components, the principles of mechanical engineering, and the physics of friction and interlocking.

At the heart of every zipper are two parallel rows of interlocking teeth or coils, usually made of metal, plastic, or nylon. When the zipper is closed, these teeth mesh together to form a continuous chain, creating a barrier that prevents the zipper from opening unintentionally. The design of the teeth varies depending on the type of zipper, with options such as metal teeth for durability and strength, plastic teeth for lightweight applications, and coil zippers for flexibility and versatility.

The slider is the component of the zipper that facilitates opening and closing by engaging and disengaging the interlocking teeth. It consists of a body or handle and two small metal or plastic tabs known as pullers, which grip the teeth and move them along the zipper tape. The slider is typically equipped with a spring-loaded mechanism that applies pressure to the teeth, ensuring a tight seal when the zipper is closed and preventing it from coming undone accidentally.

The zipper tape is the fabric strip to which the interlocking teeth are attached, forming the main body of the zipper. The tape is usually made of woven polyester or nylon and serves to anchor the teeth securely in place while providing flexibility and durability. The tape is sewn or heat-bonded to the edges of the fabric opening, creating a seamless closure that blends seamlessly with the surrounding material.

The operation of zippers relies on the principles of friction and interlocking, which govern how the teeth engage and disengage as the slider moves along the zipper tape. When the slider is pulled in one direction, the pullers grip the teeth and push them together, causing them to interlock and form a continuous chain. This process creates a tight seal that holds the fabric edges together securely, preventing gaps or openings.

Conversely, when the slider is pulled in the opposite direction, the pullers disengage from the teeth, allowing them to separate and open the zipper. This action releases the tension on the teeth, allowing the fabric edges to separate and providing access to the enclosed space. The smooth operation of the slider is facilitated by the lubricating properties of the zipper teeth, which reduce friction and ensure effortless movement along the zipper tape.

Zippers come in a variety of designs and configurations to suit different applications and preferences. Common types of zippers include coil zippers, which feature interlocking coils instead of teeth for flexibility and smooth operation, and invisible zippers, which have concealed teeth for a seamless appearance. Other variations include waterproof zippers, two-way zippers, and separating zippers, each offering unique features and benefits for specific use cases.

Chapter 40: The Principles of Parachutes: Air Resistance and Descent

Parachutes represent a remarkable application of aerodynamics and engineering principles, enabling controlled descent and safe landing for objects and people from great heights. Understanding the principles of parachutes involves exploring the dynamics of air resistance, the forces acting on a falling object, and the design considerations that govern parachute performance.

At the core of parachute operation is the principle of aerodynamic drag, which is the force exerted by air molecules on a moving object. When a parachute is deployed, it encounters air resistance as it descends through the atmosphere, causing it to slow down and decelerate. The magnitude of aerodynamic drag depends on factors such as the size and shape of the parachute canopy, the air density, the velocity of descent, and the orientation of the parachute relative to the airflow.

Parachute canopies are typically large, flat surfaces made of lightweight, durable materials such as nylon or ripstop fabric. The canopy design is optimized to maximize aerodynamic drag while minimizing weight and complexity, allowing for efficient descent and stable flight characteristics. Parachutes come in various shapes and configurations, including round, square, and ram-air canopies, each tailored to specific applications and performance requirements.

Round parachutes, also known as traditional or military parachutes, feature a simple circular design with a single layer of fabric stretched between radial lines or suspension lines. These parachutes provide reliable and predictable descent characteristics, making them suitable for cargo drops, emergency egress, and military operations. Square parachutes, on the other hand, have a more complex design with multiple chambers or cells that inflate like airfoils to generate lift and forward motion. Square parachutes offer greater maneuverability and control, making them popular choices for sport parachuting, skydiving, and aerobatic maneuvers.

Ram-air parachutes, also known as parafoils or rectangular parachutes, combine the characteristics of round and square parachutes to achieve a balance of stability, performance, and versatility. These parachutes feature a series of cells or chambers inflated by ram-air pressure, creating a rigid wing shape that provides lift and stability during descent. Ram-air parachutes are widely used in recreational skydiving, parachuting competitions, and military airborne operations.

The deployment of a parachute is a critical phase of the descent process, requiring precise timing and control to ensure safe and effective operation. Parachutes are typically deployed using one of two methods: pilot chute deployment or static line deployment. In pilot chute deployment, a small auxiliary parachute known as a pilot chute is deployed first, pulling the main parachute out of its container and allowing it to inflate and deploy fully. In static line deployment, the parachute is attached to a fixed line or static line inside the aircraft, which automatically deploys the parachute as the jumper exits the aircraft.

As a parachute descends through the atmosphere, it experiences several forces that act on it and influence its motion. The primary force acting on the parachute is gravity, which causes it to accelerate towards the Earth's surface. Aerodynamic drag, generated by the parachute canopy as it interacts with the airflow, opposes the force of gravity and slows down the descent rate. Additional forces such as lift, induced drag, and side forces may also come into play, depending on the design and orientation of the parachute.

The descent characteristics of a parachute are determined by factors such as the size and shape of the canopy, the weight of the payload, the altitude of deployment, and environmental conditions such as wind speed and direction. Engineers and aerodynamicists use mathematical models, computer simulations, and wind tunnel testing to optimize parachute designs and predict their performance under various conditions. By adjusting parameters such as canopy size, suspension line length, and deployment sequence, designers can tailor parachutes to specific mission requirements and ensure safe and efficient descent.

Chapter 41: The Strength of Bridges: Arches and Trusses

Bridges stand as iconic symbols of human ingenuity, spanning rivers, valleys, and canyons to connect communities and facilitate transportation and commerce. At the core of their structural integrity lies a careful balance of forces and materials, with arches and trusses serving as two fundamental engineering principles employed to bear loads and distribute weight efficiently. Understanding the strength of bridges involves delving into the mechanics of arches and trusses, the properties of construction materials, and the design considerations that govern their construction. This comprehensive exploration will examine the principles of arches and trusses in bridge engineering, covering topics such as their structural mechanics, historical significance, and modern applications in bridge design.

Arches are one of the oldest and most enduring structural forms used in bridge construction, dating back to ancient civilizations such as the Romans and Mesopotamians. An arch is a curved structure that spans an opening and supports the weight above it by transferring the load outward along its curved profile to its abutments or supports. This distribution of forces creates compressive stresses within the arch, which are balanced by tensile stresses in the supporting abutments, resulting in a stable and self-supporting structure.

The strength of an arch bridge lies in its ability to convert vertical loads into horizontal thrust forces, which are absorbed and resisted by the abutments at each end of the bridge. The shape and geometry of the arch determine its load-bearing capacity and stability, with designs ranging from simple semicircular arches to more complex segmented, pointed, or skewed arch configurations. Key considerations in arch bridge design include the span length, rise height, curvature, and material properties, which must be carefully optimized to ensure structural integrity and safety.

Trusses are another common structural element used in bridge construction, characterized by their triangular framework of beams or members arranged to form a rigid and stable structure. A truss bridge consists of a series of interconnected triangles that distribute loads efficiently and resist bending and torsional forces. The triangular shape of trusses provides inherent strength and stability, making them ideal for spanning long distances and supporting heavy loads with minimal material usage.

The strength of a truss bridge stems from the triangular arrangement of its members, which work together to form a rigid and lightweight structure. The top chords of the truss bear compressive forces, while the bottom chords carry tensile forces, creating a balanced system of internal stresses that resist deformation and maintain stability under load. Truss bridges come in various configurations, including Pratt, Warren, Howe, and K trusses, each offering unique advantages in terms of span length, load capacity, and construction complexity.

In addition to their inherent strength, arches and trusses offer several advantages in bridge design, including versatility, durability, and aesthetic appeal. Arches can span wide openings with minimal material usage and provide a visually striking silhouette that complements the surrounding landscape. Trusses, on the other hand, offer flexibility in span length and configuration, allowing for efficient use of materials and cost-effective construction.

The strength of bridges is further enhanced by advancements in construction materials and engineering techniques, such as reinforced concrete, steel, and composite materials. These modern materials offer superior strength, durability, and corrosion resistance compared to traditional building materials, allowing for the construction of longer, taller, and more innovative bridge designs. Computer-aided design (CAD) software and finite element analysis (FEA) tools enable engineers to simulate and optimize bridge designs, ensuring optimal performance and safety under various loading conditions.

Chapter 42: Thermometers: Measuring Temperature Accurately

Thermometers are indispensable instruments used across various fields, from meteorology and medicine to industrial processes and scientific research, to measure temperature accurately and reliably. Understanding thermometers involves exploring the principles of temperature measurement, the types of thermometers available, and the factors that influence their accuracy and precision.

Temperature is a fundamental physical quantity that reflects the average kinetic energy of particles within a substance or system. Thermometers function by detecting changes in temperature and converting them into readable numerical values, allowing for precise quantification and comparison of thermal conditions. The accuracy and reliability of temperature measurements depend on factors such as the sensitivity of the thermometer, the stability of the sensor, and the calibration procedures used to validate its performance.

One of the key components of a thermometer is the temperature sensor, which detects changes in temperature and generates a corresponding electrical signal or mechanical displacement. Common types of temperature sensors include liquid-filled bulbs, bimetallic strips, thermistors, thermocouples, and infrared sensors, each with its unique principles of operation, range of measurement, and sensitivity to temperature variations. Thermometers may also feature additional components such as digital displays, data logging capabilities, and wireless connectivity for remote monitoring and control.

The accuracy of a thermometer depends on its calibration, which involves comparing its readings to known reference standards or traceable calibration sources. Calibration ensures that the thermometer provides accurate and consistent measurements over its operating range and compensates for any systematic errors or drift in its performance. Calibration procedures may involve adjusting the thermometer's scale or offset, verifying its linearity and repeatability, and performing stability tests under controlled environmental conditions.

Thermometers are calibrated using standardized methods and equipment traceable to national or international metrology institutes, such as the National Institute of Standards and Technology (NIST) in the United States or the International Bureau of Weights and Measures (BIPM) in France. Calibration laboratories employ precision instruments such as calibrated baths, dry-well calibrators, and reference thermometers to establish traceable measurement chains and ensure the accuracy and reliability of temperature measurements.

Temperature measurement is typically expressed in one of several temperature scales, including Celsius (°C), Fahrenheit (°F), and Kelvin (K), each with its zero point and unit interval defined based on specific reference points or physical phenomena. The Celsius scale is based on the freezing and boiling points of water at standard atmospheric pressure, with 0°C representing the freezing point and 100°C representing the boiling point. The Fahrenheit scale, used primarily in the United States, defines 32°F as the freezing point and 212°F as the boiling point of water under standard conditions. The Kelvin scale, which is the standard unit of temperature in the International System of Units (SI), is based on the absolute zero point, where all thermal motion ceases, and defines 0 K as absolute zero.

Thermometers find applications in a wide range of industries and environments, from monitoring environmental conditions in weather stations and climate control systems to ensuring product quality and safety in food processing, pharmaceutical manufacturing, and healthcare facilities. In meteorology, thermometers are used to measure air temperature, humidity, and dew point, providing essential data for weather forecasting, climate studies, and environmental monitoring. In medical settings, thermometers are used to measure body temperature accurately, helping diagnose and monitor various health conditions and diseases.

In industrial processes, thermometers play a critical role in monitoring and controlling temperature-sensitive processes such as chemical reactions, heat treatment, and manufacturing operations. Thermocouples and resistance temperature detectors (RTDs) are commonly used in industrial applications due to their ruggedness, reliability, and wide temperature range. Infrared thermometers and thermal imaging cameras are also used to measure surface temperatures remotely and detect hot spots in electrical systems, mechanical equipment, and building structures.

Chapter 43: The Aerodynamics of Cars: Streamlining for Speed

The aerodynamics of cars is a fascinating field that encompasses the study of airflow around vehicles and its impact on performance, efficiency, and handling. Understanding car aerodynamics involves exploring concepts such as drag, lift, downforce, and streamlining, as well as the design principles and technologies used to optimize airflow and reduce aerodynamic resistance.

Aerodynamic drag is the force exerted by air resistance on a moving vehicle, which acts to oppose its motion and reduce speed and fuel efficiency. Drag is caused by the interaction between the vehicle's shape and the surrounding airflow, resulting in turbulent eddies and pressure variations that create drag forces. The key to reducing aerodynamic drag is to minimize the vehicle's frontal area, streamline its shape, and optimize airflow around critical areas such as the body, wheels, and undercarriage.

Streamlining is a design principle used to reduce aerodynamic drag by shaping the vehicle's body to minimize air resistance and turbulence. Streamlined shapes such as teardrops, airfoils, and ellipsoids are characterized by smooth, curved surfaces that allow air to flow smoothly around the vehicle, minimizing drag and reducing fuel consumption. Streamlining also reduces noise, vibration, and wind noise, improving comfort and stability at high speeds.

Aerodynamic devices such as spoilers, diffusers, and splitters are used to modify airflow around the vehicle and enhance its aerodynamic performance. Spoilers are airfoils mounted on the rear of the vehicle to reduce lift and increase stability by redirecting airflow downward. Diffusers are aerodynamic surfaces located under the rear of the vehicle, which accelerate airflow and create low-pressure zones to reduce drag. Splitters are horizontal fins mounted on the front of the vehicle to direct airflow around the sides and under the car, reducing lift and improving downforce.

Wind tunnel testing is a crucial tool used in car aerodynamics to evaluate the vehicle's aerodynamic performance and optimize its design. Wind tunnels simulate real-world airflow conditions by generating controlled airflows around scale models or full-size vehicles, allowing engineers to measure aerodynamic forces, pressure distributions, and flow patterns. Wind tunnel testing enables designers to identify areas of high drag, turbulence, and lift and develop solutions to improve aerodynamic efficiency and performance.

The role of aerodynamics in vehicle design and performance is significant, particularly in high-performance cars and racing vehicles where aerodynamic performance directly affects speed, handling, and cornering ability. Racing cars are designed with aerodynamic features such as spoilers, diffusers, and aerodynamic body shapes to maximize downforce and minimize drag, improving traction and stability at high speeds. Formula 1 cars, for example, use advanced aerodynamic designs to generate massive amounts of downforce, allowing them to corner at high speeds and maintain grip in challenging conditions.

In addition to performance benefits, aerodynamics also plays a crucial role in vehicle safety, stability, and fuel efficiency. Cars with aerodynamic designs experience less wind resistance, resulting in improved fuel economy and reduced emissions. Aerodynamic stability is also essential for vehicle safety, particularly at high speeds, where aerodynamic forces can affect vehicle handling and stability. Advanced aerodynamic features such as active aerodynamics, adaptive spoilers, and variable air intakes are increasingly used in modern cars to optimize aerodynamic performance and improve overall efficiency and performance.

Chapter 44: Helmets: Safety through Impact Resistance

Helmets are critical safety equipment designed to protect the head from impacts and reduce the risk of traumatic brain injuries in a wide range of activities, including sports, cycling, motorcycling, and industrial work. Understanding helmets involves exploring the principles of impact resistance, helmet design, materials engineering, and safety standards, as well as the role of helmets in preventing head injuries and saving lives.

The primary function of a helmet is to absorb and dissipate the energy from an impact, reducing the forces transmitted to the wearer's head and minimizing the risk of injury. Helmets achieve this through a combination of design features, materials, and construction methods that provide impact resistance and structural integrity. The outer shell of the helmet is typically made of tough, impact-resistant materials such as polycarbonate, fiberglass, or carbon fiber, which provide a protective barrier against external forces and distribute impact energy over a larger area.

Inside the helmet shell is a layer of energy-absorbing material, usually foam or expanded polystyrene (EPS), which compresses upon impact to absorb and dissipate kinetic energy. This crushable foam layer acts as a cushion, decelerating the head and reducing the severity of acceleration forces experienced during a collision. The thickness, density, and compression characteristics of the foam liner are carefully engineered to optimize impact absorption and protect against head injuries across a range of impact speeds and conditions.

Helmet design also incorporates features such as ventilation channels, padding, and retention systems to enhance comfort, fit, and stability while maximizing impact protection. Ventilation openings and channels allow airflow to circulate inside the helmet, helping to dissipate heat and moisture and keep the wearer cool and comfortable during prolonged use. Soft padding materials such as foam or gel provide cushioning and support, while adjustable retention systems such as straps and buckles allow for a customized fit and secure helmet positioning on the head.

One of the critical aspects of helmet safety is certification and compliance with industry standards and testing protocols. Regulatory organizations such as the Snell Memorial Foundation, the American National Standards Institute (ANSI), and the European Committee for Standardization (CEN) establish minimum performance requirements and testing procedures for helmets used in various activities and industries. These standards specify criteria such as impact energy attenuation, penetration resistance, strap strength, and peripheral vision to ensure helmets provide adequate protection against common hazards and injury mechanisms.

Helmet testing involves subjecting prototypes or production samples to controlled impact simulations and environmental conditions to evaluate their performance under realistic scenarios. Common testing methods include drop tests, where helmets are dropped from specified heights onto various surfaces to measure impact absorption, and penetration tests, where sharp objects are applied to the helmet surface to assess puncture resistance. Helmets must meet or exceed specified performance thresholds to receive certification and approval for sale and use in the intended application.

Advancements in helmet technology continue to improve safety and performance, with innovations such as multi-impact foams, rotational impact protection systems (RIPS), and custom-fit 3D-printed helmets pushing the boundaries of helmet design and engineering. Multi-impact foams are engineered to withstand repeated impacts without losing their protective properties, making them ideal for high-risk sports such as skateboarding and snowboarding. RIPS systems incorporate sliding or rotational elements into the helmet design to mitigate the effects of angular impacts and reduce the risk of rotational brain injuries, which are common in certain sports such as football and cycling.

Custom-fit 3D-printed helmets use advanced scanning and manufacturing technologies to create personalized helmets tailored to the wearer's head shape and size. These helmets offer superior fit, comfort, and protection compared to traditional off-the-shelf helmets, reducing pressure points, improving ventilation, and enhancing overall safety and performance. Custom-fit helmets are increasingly used in professional sports, military applications, and industrial settings where precise fit and maximum protection are critical.

Chapter 45: The Physics of Ping Pong: Spin and Momentum

The physics of ping pong, also known as table tennis, is a captivating blend of mechanics, aerodynamics, and strategy, where players manipulate spin, speed, and trajectory to outmaneuver opponents and win points. Understanding the physics of ping pong involves exploring concepts such as spin, momentum, ball trajectory, and paddle dynamics, as well as the interactions between these factors that determine the outcome of each shot.

Spin is a fundamental aspect of ping pong physics that significantly influences ball behavior and player tactics. Spin is imparted to the ball through the rotation of the paddle during contact, causing the ball to spin in a clockwise or counterclockwise direction as it travels through the air. Topspin, backspin, sidespin, and no-spin are the primary spin types used in ping pong, each producing distinct flight paths and bounce characteristics that players exploit to control the game.

Topspin is a forward rotation of the ball, generated by brushing the paddle upward and forward at the moment of impact. Topspin causes the ball to dip downward quickly after crossing the net, making it difficult for opponents to return with accuracy and consistency. Backspin, on the other hand, is a backward rotation of the ball, created by brushing the paddle downward and backward during contact. Backspin produces a floaty trajectory and a low bounce, making it challenging for opponents to generate power and pace on their shots.

Sidespin is a lateral rotation of the ball, generated by brushing the paddle across the side of the ball at an angle. Sidespin causes the ball to curve in flight and bounce unpredictably off the opponent's paddle, creating opportunities for deception and misdirection. No-spin shots, also known as dead balls, involve minimal or no rotation of the ball, resulting in a flat trajectory and bounce that catches opponents off guard and disrupts their timing and rhythm.

The interaction between spin and air resistance, known as the Magnus effect, further influences ball behavior in ping pong. The Magnus effect causes spinning objects such as ping pong balls to deviate from their expected trajectory due to differences in air pressure and flow around the ball's surface. Topspin produces a downward force that enhances ball stability and control, while backspin creates an upward force that slows the ball's descent and increases hang time.

Ball bounce is another critical aspect of ping pong physics that affects shot placement, timing, and strategy. The bounce height, speed, and angle of the ball after contacting the table depend on factors such as spin, speed, and angle of incidence. Topspin shots tend to bounce low and fast off the table, while backspin shots bounce high and slow, making them easier to return with precision. Sidespin shots can produce unpredictable bounces and angles, challenging opponents to anticipate and adjust their positioning and footwork.

Paddle design and materials also play a significant role in ping pong physics, influencing shot control, power generation, and spin manipulation. Modern ping pong paddles feature a combination of rubber, sponge, and blade materials optimized for speed, spin, and control. The rubber surface of the paddle, known as the rubber sheet, comes in various textures and thicknesses, each offering unique playing characteristics such as tackiness, grip, and spin sensitivity.

The sponge layer underneath the rubber sheet acts as a shock absorber, dampening vibrations and providing additional power and speed to shots. The blade or handle of the paddle is typically made of wood or composite materials and contributes to the paddle's stiffness, weight distribution, and balance. Players customize their paddles to suit their playing style and preferences, selecting rubbers, sponges, and blades that optimize spin, speed, and control for their individual needs.

The physics of ping pong extends beyond shot mechanics to encompass player technique, footwork, and strategy, where players leverage physics principles to gain a competitive edge. Advanced players use spin, speed, and placement to exploit opponents' weaknesses, force errors, and control the pace and flow of the game. Tactics such as looping, chopping, smashing, and blocking involve precise timing, coordination, and anticipation of spin and trajectory to execute effectively and outmaneuver opponents.

Chapter 46: How Soap Cleans: Surface Tension and Micelles

Understanding how soap cleans involves delving into the intricate chemistry of surfactants, micelles, and surface tension. Soap is a surfactant, a compound that lowers the surface tension of water, allowing it to spread and penetrate into crevices and pores to lift dirt, oil, and grease from surfaces. This detailed exploration will cover the molecular mechanisms of soap action, the formation of micelles, and the role of surface tension in cleaning processes.

At the heart of soap's cleaning power lies its unique molecular structure, which consists of a hydrophilic (water-attracting) head and a hydrophobic (water-repelling) tail. This amphiphilic nature allows soap molecules to interact with both water and oily substances, facilitating the emulsification and removal of dirt and grease from surfaces. When soap is added to water, its hydrophilic heads orient themselves towards the water molecules, while the hydrophobic tails cluster together to form micelles.

Micelles are spherical aggregates of soap molecules arranged with their hydrophilic heads facing outward and their hydrophobic tails shielded inside the micelle. This structure allows micelles to solubilize hydrophobic compounds such as oil and grease by encapsulating them within the micelle core, effectively trapping them in a water-soluble shell. As a result, the dirt and grease adhering to surfaces are enveloped by micelles and lifted away from the surface, where they can be rinsed away with water.

The formation of micelles is driven by the need to minimize the free energy of the system, as soap molecules self-assemble to maximize their interactions with water and minimize their interactions with air or oily substances. When soap is added to water, it disrupts the intermolecular forces between water molecules, reducing the surface tension and allowing water to wet surfaces more effectively. This reduction in surface tension enables water and soap to penetrate into pores and crevices, loosening and dislodging dirt and debris from surfaces.

The cleaning action of soap relies on a combination of physical and chemical processes, including emulsification, dispersion, and dissolution. Emulsification occurs when soap molecules surround and suspend oil droplets in water, preventing them from re-aggregating and redepositing on surfaces. Dispersion involves breaking up large particles of dirt or grease into smaller, more manageable fragments that can be dispersed and rinsed away with water. Dissolution occurs when soap molecules solubilize oil and grease, allowing them to mix with water and be carried away in solution.

Surface tension plays a crucial role in soap's cleaning efficacy by enabling water to spread and penetrate into small spaces and tight corners, where dirt and grease may accumulate. By reducing the surface tension of water, soap increases its wetting ability and capillary action, allowing it to flow more freely and access hard-to-reach areas. This enhanced wetting and spreading action helps displace and dissolve dirt and grease from surfaces, making them easier to clean and maintain.

In addition to its cleaning properties, soap also possesses antimicrobial properties that help kill or inhibit the growth of bacteria and viruses on surfaces. The hydrophobic tails of soap molecules can disrupt the lipid membranes of bacteria and enveloped viruses, causing them to lyse or lose their structural integrity. This antimicrobial action, combined with soap's cleaning and emulsifying properties, makes it an effective tool for maintaining hygiene and preventing the spread of infectious diseases.

Chapter 47: Wristwatches: Timekeeping on Your Wrist

Wristwatches represent a fascinating intersection of horology, engineering, and fashion, offering a convenient and stylish way to keep track of time while also serving as a symbol of personal style and status. Understanding wristwatches involves exploring their history, mechanics, technology, and cultural significance, as well as the innovations and trends that have shaped their evolution over time. This detailed exploration will cover topics such as the origins of wristwatches, the mechanics of timekeeping, the components of a watch, and the role of wristwatches in modern society.

The history of wristwatches dates back to the late 19th century when pocket watches were adapted for wear on the wrist for military and practical purposes. Early wristwatches, known as "wristlets" or "trench watches," were typically worn by soldiers in the field to facilitate timekeeping during combat operations. These early wristwatches featured rugged construction, simple designs, and manual winding mechanisms, making them reliable timepieces for military use.

Over time, wristwatches evolved from functional timekeeping devices into fashionable accessories and status symbols, with manufacturers incorporating innovative designs, materials, and features to appeal to a broader audience. The introduction of automatic (self-winding) movements, waterproof cases, and shock-resistant designs further enhanced the durability, reliability, and versatility of wristwatches, making them suitable for everyday wear in various environments and activities.

The mechanics of timekeeping in wristwatches involve a complex interplay of gears, springs, escapements, and oscillators that work together to measure and display time accurately. The heart of a mechanical watch is its movement, which comprises hundreds of precision-engineered components assembled with meticulous craftsmanship. Mechanical movements are powered by a mainspring, which stores energy when wound and releases it gradually to drive the gears and regulate the motion of the hands.

The escapement mechanism, typically a lever or wheel, controls the release of energy from the mainspring and ensures that the movement of the hands is regulated and consistent. The balance wheel, oscillating back and forth at a precise frequency, acts as the timekeeping element of the movement, dividing time into equal intervals and driving the motion of the hands. The escapement mechanism and balance wheel are housed within the watch's movement, protected by a series of gears and bridges that transmit power and motion throughout the watch.

In addition to mechanical movements, wristwatches may also feature electronic quartz movements, which utilize a small quartz crystal to regulate the passage of time. Quartz watches are powered by a battery that sends electrical impulses to the quartz crystal, causing it to vibrate at a precise frequency. These vibrations are then converted into electrical signals that drive a stepper motor, which in turn moves the hands of the watch. Quartz movements are known for their accuracy, reliability, and affordability, making them popular choices for mass-produced wristwatches.

The components of a wristwatch are housed within a case, typically made of metal, plastic, or ceramic, which protects the movement from dust, moisture, and impact damage. The case may feature a crystal, usually made of glass or synthetic sapphire, which covers the dial and hands of the watch and provides visibility and protection against scratches and abrasions. The dial, or face of the watch, contains the hour markers, numerals, and hands that indicate the time, as well as additional features such as chronograph subdials, date windows, and moon phase displays.

The strap or bracelet of a wristwatch is used to secure the watch to the wearer's wrist and may be made of leather, metal, rubber, or fabric, depending on the style and function of the watch. Straps and bracelets come in various designs, colors, and materials, allowing for customization and personalization to suit individual preferences and occasions. Some wristwatches feature interchangeable straps or bracelets, allowing wearers to change the look and feel of their watch to match their outfit or mood.

The role of wristwatches in modern society extends beyond timekeeping to encompass fashion, identity, and self-expression, with wristwatches serving as statements of personal style, taste, and status. Luxury wristwatch brands such as Rolex, Patek Philippe, and Audemars Piguet are revered for their heritage, craftsmanship, and exclusivity, with collectors and enthusiasts worldwide seeking out rare and limited-edition models for their investment value and aesthetic appeal.

Chapter 48: Windshield Wipers: Oscillation and Motion

Windshield wipers are essential automotive accessories designed to keep windshields clear of rain, snow, dirt, and debris, ensuring visibility and safety for drivers in various weather conditions. Understanding windshield wipers involves exploring their mechanics, motion, design, and technological advancements, as well as their role in maintaining visibility and enhancing driving comfort. This comprehensive exploration will cover topics such as the history of windshield wipers, the principles of oscillation and motion, wiper blade design, and innovations in wiper technology.

The history of windshield wipers dates back to the early 20th century when motorists began seeking solutions to clear rain and snow from their windshields for improved visibility while driving. Early windshield wipers were manually operated, consisting of rubber blades attached to mechanical arms that were operated by hand or foot. These early wipers were primitive compared to modern systems but represented a significant advancement in automotive safety and convenience.

Over time, windshield wiper systems evolved to incorporate electric motors, linkage mechanisms, and intermittent wiping modes to provide more reliable and efficient operation. Electric windshield wipers, first introduced in the 1920s, used electric motors to drive the wiper arms and blades, eliminating the need for manual operation. Linkage mechanisms, such as scissor arms and pantographs, were developed to convert the rotational motion of the wiper motor into the linear motion required to move the wiper blades across the windshield.

The principles of oscillation and motion govern the operation of windshield wipers, enabling them to sweep across the windshield in a controlled and efficient manner. Oscillation refers to the back-and-forth movement of the wiper blades, which is essential for clearing water, snow, and debris from the windshield surface. The motion of the wiper blades is driven by the wiper motor, which generates rotational force transferred to the wiper arms and blades through a series of linkages and pivots.

The design of windshield wiper blades plays a crucial role in their effectiveness and performance, with modern blades featuring advanced materials, aerodynamic profiles, and pressure distribution systems to optimize contact with the windshield surface. Wiper blades are typically made of rubber or silicone compounds that provide flexibility, durability, and resistance to environmental factors such as UV radiation, heat, and cold. The shape and contour of the wiper blades are engineered to conform to the curvature of the windshield and maintain consistent pressure across the entire wiping area.

Innovations in windshield wiper technology have led to the development of features such as rain-sensing systems, heated wiper blades, and adaptive wiping modes that enhance visibility and convenience for drivers. Rain-sensing wiper systems use sensors to detect moisture on the windshield and automatically activate the wipers at the appropriate speed and interval to maintain clear visibility. Heated wiper blades incorporate heating elements into the blade structure to prevent ice and snow buildup and ensure uninterrupted wiping in cold weather conditions. Adaptive wiping modes adjust the speed and frequency of the wiper blades based on driving speed, vehicle speed, and environmental conditions to optimize wiping performance and minimize noise and vibration.

Windshield wipers play a critical role in maintaining visibility and safety for drivers, particularly in adverse weather conditions such as rain, snow, fog, and sleet. Proper maintenance and care of windshield wipers are essential to ensure their effectiveness and longevity, including regular inspection, cleaning, and replacement of worn or damaged blades. Drivers should also use windshield washer fluid to lubricate and clean the windshield surface, allowing wipers to operate smoothly and clear debris effectively.

Chapter 49: Smart Home Devices: IoT and Automation

Smart home devices represent the convergence of technology and everyday living, offering homeowners convenience, efficiency, and control over their living spaces through interconnected devices and automation. Understanding smart home devices involves exploring the Internet of Things (IoT), wireless communication protocols, device ecosystems, and the practical applications of automation in residential settings. This detailed exploration will cover topics such as the evolution of smart home technology, the principles of IoT connectivity, popular smart home devices, and the benefits and challenges of home automation.

The evolution of smart home technology can be traced back to the early 2000s when the first wave of interconnected devices and home automation systems began to emerge. These early systems were often complex, expensive, and limited in functionality, requiring professional installation and programming. However, advancements in wireless communication, sensor technology, and cloud computing have since transformed the smart home landscape, making it more accessible, affordable, and user-friendly for consumers.

At the heart of smart home technology is the Internet of Things (IoT), a network of interconnected devices and sensors that communicate and exchange data over the internet. IoT enables smart home devices to connect, interact, and coordinate their functions seamlessly, allowing users to monitor and control various aspects of their homes remotely via smartphones, tablets, or voice assistants. IoT connectivity relies on wireless communication protocols such as Wi-Fi, Bluetooth, Zigbee, and Z-Wave, which enable devices to communicate with each other and with cloud-based platforms and services.

Smart home devices encompass a wide range of products and systems designed to automate and enhance various aspects of home life, including lighting, heating, security, entertainment, and appliances. Popular smart home devices include smart thermostats, which allow users to control heating and cooling systems remotely and adjust temperature settings based on occupancy and preferences. Smart lighting systems enable users to customize lighting schedules, brightness levels, and colors using mobile apps or voice commands, while smart locks provide keyless entry and remote access control for enhanced security and convenience.

Home automation platforms and ecosystems, such as Amazon Alexa, Google Assistant, Apple HomeKit, and Samsung SmartThings, serve as central hubs for managing and integrating smart home devices from different manufacturers. These platforms offer compatibility with a wide range of devices and services, allowing users to create custom automation routines, scenes, and triggers to streamline daily tasks and routines. For example, users can create "good morning" routines that turn on lights, adjust thermostat settings, and brew coffee automatically upon waking up.

The benefits of smart home devices and automation are manifold, ranging from convenience and energy savings to improved security and comfort. Smart thermostats, for instance, can help homeowners save energy and reduce utility bills by automatically adjusting temperature settings based on occupancy patterns and weather conditions. Smart security cameras and doorbell cameras provide real-time video monitoring and alerts, allowing users to keep an eye on their homes and deter potential intruders from anywhere with an internet connection.

In addition to convenience and security, smart home devices can also enhance comfort and accessibility for users with disabilities or mobility limitations. Voice-controlled assistants such as Amazon Alexa and

Google Assistant enable hands-free operation of smart devices and services, making it easier for users to control lights, appliances, and entertainment systems without physical interaction. Smart home automation can also provide peace of mind for caregivers and family members by remotely monitoring the well-being and activities of elderly or vulnerable individuals.

Despite the many benefits of smart home devices and automation, there are also challenges and considerations that homeowners should be aware of, including privacy concerns, interoperability issues, and security risks. Smart home devices collect and transmit data about users' habits, preferences, and activities, raising concerns about data privacy and potential misuse of personal information. Interoperability issues may arise when integrating devices from different manufacturers or ecosystems, requiring users to choose compatible products or invest in third-party bridges or hubs to enable communication between devices.

Security risks associated with smart home devices include vulnerabilities in device firmware, cloud services, and communication protocols that could be exploited by hackers or malicious actors to gain unauthorized access to homes or compromise user privacy. To mitigate these risks, homeowners should follow best practices for securing their smart home networks, such as using strong passwords, keeping firmware and software up to date, and configuring privacy settings and access controls appropriately.

Chapter 50: Conclusion: Embracing Physics in Everyday Life

Embracing physics in everyday life means recognizing and appreciating the fundamental principles of the physical world that shape our experiences, interactions, and environments. From the natural forces governing celestial bodies to the intricate mechanisms behind everyday objects, physics permeates every aspect of our existence, offering insights into the workings of the universe and empowering us to understand, manipulate, and harness the forces of nature for practical purposes.

Throughout this exploration of the science behind common objects and phenomena, we've seen how physics influences and informs the design, function, and behavior of the world around us. Whether it's the mechanics of motion in bicycles, the dynamics of fluid flow in faucets, or the principles of electromagnetism in smartphones, physics provides a framework for understanding the underlying mechanisms and phenomena that govern our daily lives.

By understanding the principles of physics, we gain a deeper appreciation for the beauty, complexity, and elegance of the natural world, from the graceful motion of celestial bodies to the intricate structures of atoms and molecules. Physics invites us to explore the mysteries of the cosmos, from the smallest subatomic particles to the vast expanse of the universe, and to marvel at the interconnectedness of all things.

Moreover, embracing physics in everyday life empowers us to become more informed and engaged citizens, capable of making informed decisions about technology, policy, and the environment. Whether it's evaluating the energy efficiency of household appliances, understanding the environmental impact of transportation options, or advocating for evidence-based policies to address climate change, a basic understanding of physics equips us with the knowledge and critical thinking skills needed to navigate complex issues and contribute to positive change.

Furthermore, physics serves as a catalyst for innovation and progress, driving advancements in science, technology, engineering, and medicine that improve our quality of life and expand our understanding of the universe. From the invention of the light bulb to the development of quantum mechanics, physics has fueled countless breakthroughs and discoveries that have revolutionized the way we live, work, and communicate.

Epilogue

As we reach the conclusion of our exploration in "The Science Behind Common Objects: Grasping the Physics in Our Daily Lives," it is appropriate to contemplate the knowledge gained, the breakthroughs achieved, and the marvels encountered along the way. We have delved into the depths of physics, investigating the intricate mechanisms and phenomena that shape the behavior of the objects we encounter in our everyday lives, ranging from the ordinary to the extraordinary.

Throughout our journey, we have developed a newfound appreciation for the elegance and intricacy of the physical world. We have acquired an understanding of how the principles of physics govern the movement of a bicycle, the functioning of a faucet, the illumination of a light bulb, and the operation of countless other objects that populate our existence. We have been astounded by the ingenuity of human invention, the beauty of mathematical relationships, and the interconnection between science and our daily routines.

However, beyond the scientific knowledge acquired, our expedition has been one of awe, inquisitiveness, and revelation. We have caught a glimpse of the poetry of motion in the graceful swing of a pendulum, experienced the impact of innovation in the click of a door handle, and witnessed the transformative potential of technology in the glow of a smartphone screen.

As we bid farewell to the objects that have accompanied us on this voyage, let us carry with us a deeper comprehension of the world that surrounds us and a renewed sense of wonder for the mysteries that still lie undiscovered. Let us continue to embrace the spirit of curiosity and exploration that has guided us on this adventure, and let us never lose sight of the profound beauty and complexity of the universe in which we reside.

As we bring this book to a close, it is important to bear in mind that our world is filled with hidden marvels just waiting to be discovered. The quest for knowledge is a never-ending voyage, and it is my hope that our inquisitiveness will continue to guide us, our sense of awe will continue to motivate us, and our spirit of exploration will continue to enhance our lives in the future.

Thank you for joining me on this journey, and may your own explorations of the world around you be filled with discovery, insight, and joy.

The End.

www.ingramcontent.com/pod-product-compliance
Lightning Source LLC
Chambersburg PA
CBHW020339160726
47992CB00004B/1891